THE NATURE-WATCHERS

THE NATURE WATCHERS

Exploring Wildlife with the Experts

ROBIN BROWN
and JULIAN PETTIFER

COLLINS
8 Grafton Street, London W1
1985

Also by Robin Brown

WHEN THE WOOD BECAME THE TREES

A FOREST IS A LONG TIME GROWING

MEGALODON

THE LURE OF THE DOLPHIN

Also by Julian Pettifer

DIAMONDS IN THE SKY (with Kenneth Hudson)

AUTOMANIA (with Nigel Turner)

Also by Robin Brown and Julian Pettifer

NATURE WATCH

William Collins Sons and Co Ltd
London · Glasgow · Sydney · Auckland · Toronto ·
Johannesburg

This publication is based on the television series *Nature Watch*
produced by Central Independent Television plc

CENTRAL
Central logo copyright © 1982
Central Independent Television plc

First published 1985
© Robin Brown and Roland Productions Limited 1985

ISBN 0 00 219149 0

Typeset by Ace Filmsetting Limited, Frome
Colour originated by Newsele srl (UK), Milan, Italy
Printed and bound by
New Interlitho SpA, Milan, Italy

CONTENTS

Our thanks to all our subjects who guided us through a vast library of natural history, and to those who helped us contain the essence of it in these pages: in particular Amanda Kent, Miranda Weston-Smith, Val Noel Finch, Tish Faith, Heather Lowe and Bryony Kinnear.

INTRODUCTION

The wonder of the planet Earth is that it is teeming
with life. Human interest in nature focuses on this
kaleidoscope of living things; the less animate an
organism, the less it attracts our interest. Planets that
might sustain 'life' intrigue us, those composed of
arid crystals we view with disappointment. The
American naturalist, Loren Eisely, has aptly identified
this yearning as a search for an end to the 'long loneli-
ness' by a species which regards itself as unique but
does not know why.

This obsession with questioning how and why life
came about on Earth, and our place in the fabric, was
given the label *natural history* when it became a
science or, more accurately, when thinking human
beings dared question the Biblical explanation of
creation as a seven-day period of divine hyper-
activity. This is not to suggest that the pioneer
nature-watchers like Charles Darwin and Albert
Wallace were agnostic. The opposite is true. They
explored the world, making minute and painstaking
observations of what they saw, in the belief that the
Biblical interpretations were the innocent allegories
of a race trying to come to terms with mysteries
beyond its comprehension, and that the truth was an
equally holy grail.

Eventually in 1859 Darwin published *The Origin of
Species*, and in it he suggested that evolution occurs as
a result of natural selection rather than divine inter-
vention. He knew that this theory would 'bring down
the wrath of God' (or, as indeed happened, the wrath
of theologians), but the very act of publishing his
treatise was a statement in itself that knowledge, and
indeed morality, must evolve as well. Darwin made a
good case for thinking that to understand the nature
of life and its origins, the present is as relevant as the
past and that our essential interest in life is not just a
matter of its history but of the evolving miracle.

The label *natural history* does no justice to this
wide-ranging obsession, hence the title of this book
and the television series which is its companion. All
of us, from scientist to amateur, are engrossed in a
global, perhaps universal, nature-watch. Indeed it
appears to be a 'natural' facet of the human psyche, an
interest inherent in us all. If you are in any doubt

INTRODUCTION

about your rights to a place in this society, be assured that amateurs have always been its most important members.

Some 300 years ago, young apprentice apothecaries became the first naturalists when they went out into the country in search of useful 'herbals' growing wild. By 1689 they had formed a natural history society, the Temple House Botanic Club, and this was soon followed by more specialized societies, like the Society of Aurelians which was devoted to entomology (the study of insects).

Intellectual amateurs were the mainstay of these societies. Gilbert White (pictured here as a young ecclesiastic at Oriel College, Oxford) is regarded as the father of British natural history. He was a country parson and his classic collection *The Natural History of Selborne*, published in 1789, is the product of what he saw on local walks and described in letters to a naturalist friend. He repeatedly refused promotion in order to remain a curate watching nature in his beloved parish.

Darwin himself was also an amateur naturalist when

Below: *Nature-watching begins at home. A shallow pond, like this one at Robin Brown's home in Northamptonshire greatly enhances the number of creatures and plants that may be observed at leisure in the garden. Birds in particular enjoy this facility for drinking and bathing; herons for fishing, ducks for paddling, swallows for insect food. Frogs, newts and a host of aquatics become next-door neighbours.*

he went aboard the *Beagle* in 1831, taking up what seemed an interesting opportunity after university. On that four-year voyage, the excellent nature-watching experience he gained laid the ground for his *Origin of Species* (which did not appear until 23 years later).

So *The Nature-watchers* is dedicated to the grand tradition of natural history: to the enthusiastic amateur. It is a book by and for people in love with nature; a record of the work of the world's most famous nature-watchers and some of the least famous, of scientists and of complete amateurs.

It is a companion to our television programme *Nature Watch*, not just a recycling of that material. These pages have allowed us to include a great deal of natural history which could not be accommodated in a half-hour television programme. We are, after all, dealing with knowledge that has taken our nature-watchers a lifetime to acquire.

This volume also makes it possible to set our subjects in the broader canvas of an international nature-watch and to include studies on the three great domains of nature: the sea, the land, and the sky. Each environment is explored within the perspective of evolution. To gain as wide a perspective as possible we have included material from past and present *Nature Watch* programmes and, in one or two cases, from nature-watchers we have interviewed and will either be filming while this book is being printed or will be featuring in future programmes. We have also, reluctantly, had to leave a few people out for want of space. We have supplied information on how to acquire specialized naturalist skills (like scuba-diving, bee-keeping, the establishment of a wildlife garden, and wildlife photography) that will greatly enhance your nature-watching. We have further included, in separate 'boxes', material of a biological and zoological nature to support the text, which was not included in the television series.

This book is certainly the first collected record of work by some of the world's greatest living naturalists. We hope it will inspire you to join them, in the knowledge that you will then share their unique sense of fulfilment in what really is one of the most worthwhile of human endeavours.

The international nature-watch needs as many supporters as it can muster. Sadly it is all too often a doom-watch, and while we always attempt to point out the achievements of international conservation, the fact has to be faced that an emergency 'recovery

Above: *The father of British Natural History, Gilbert White: a rare sketch from Oriel College, Oxford, before he became the 'naturally' inquisitive vicar of Selbourne.*

operation' is well overdue. The unfortunate legacy of today's nature-watchers is several centuries of what can only be described as mindless attrition, and if there is to be any nature worth being watched in the twenty-first century, the rescue has to be mounted now, and with great energy.

In each of our chapters, therefore, you will find otherwise happy men and women sadly contemplating the future. Professor Konrad Lorenz intimated that he regarded the human race as a dangerously disturbed species, travelling to its own destruction down a blind alley of evolution while it uses up the resources of the world in pursuit of a pointless set of material values.

If we are to stem vandalism against nature we have also to resolve a dangerous dichotomy that is developing between the rich and the poor nations of the world on matters of conservation. On the one hand we have the affluent West preaching the ecological message through powerful lobbies like the World Wildlife Fund, messages which sound all too often like echoes from an old colonial club.

On the other is a Third World concerned with the much more basic issues of starvation, land hunger, subsistence economies and escaping from second-class statehood. They see great hypocrisy in the sermon from a mount that has largely been denuded of its wildlife in the interest of the West's industrial revolution.

We believe that the very least the affluent nations can do is help pay for the preservation of Third World wildlife as if we were paying rent on what is, after all, a world heritage. The Third World can provide the room, if we take care of the board, at least in the interim.

Robin Brown/Julian Pettifer 1985

SEA

1. SEA-WATCH

More than three-quarters of the Earth is covered by water. The average depth of the sea is $2\frac{1}{3}$ miles! Smooth out the bumps on land by levelling the mountains into the ravines, canyons and gorges, and the surface of the Earth would be submerged to an average depth of almost $1\frac{1}{2}$ miles. The Earth has often been described as a pebble spinning in space; it could also be said that it is a rotating puddle from which scraps of dry land protrude.

Not surprisingly this vast puddle is where most of the inhabitants of the Earth reside. It is small wonder too that the variety of creatures you may watch in the sea verges on the infinite, that it is the melting pot of all life, and that we have yet to scratch the surface of its natural history.

It was in the seas that life first began and so it is appropriate for our nature-watch to begin in this realm. Life is believed to have originated when groups of simple molecules coalesced to form compounds in the primordial seas. It is probably because they offered the best protection against the vagaries of climate and the deadly assault of cosmic radiation that life was initiated in these nutrient-rich soups. We know from the most ancient fossil records that structures resembling bacteria and single-celled, spherical, blue-green algae were alive approximately 3000 million years ago in the oceans. The Earth itself is around 4500 million years old.

Because the sea is such a stable place in which to live, many marine animals were able to avoid the need for dramatic evolutionary change and have remained almost exactly as they were at the dawn of time; they are today's living fossils. Others have never found the need to leave the sea in search of more varied environments. All the echinoderms (such as sea cucumbers, sea urchins and starfish) still, and quite literally, stick to the sea bed.

Other creatures have made the round trip, electing to return to the sea's embrace after exploratory excursions, spanning millions of years, on land. These include a group of highly intelligent animals, the whales and dolphins (see Chapter 3).

The earliest living creatures were exclusively marine. Of these the trilobites have attracted a great

What's in a Name?

Linnaeus and the system of Nature
Why are plants and animals given complicated Latin names? What is wrong with 'Pinks' rather than *Caryophyllaceae dianthus* or 'Plaice' instead of *Pleuronectes platessa*?

Somewhat ironically the naturalist reponsible for what are often regarded as unnecessarily complex (and certainly unpronounceable) names actually changed his own name. He was the Swede, Carl von Linne, who decided to be known as Linneaus.

It would be impossible to keep accurate track of the millions of organisms on earth by common names (like pink or plaice); there are simply too many fine variations. In the mid 1700s, von Linne created a filing system using Latin headings.

Linnaeus's filing system broke all living things down into five expanding categories – classes, orders, families, genera and, finally, species: with some improvements it is still the filing system in use today.

The names Linnaeus gave to organisms were descriptive filing systems in themselves, and this precedent has been carried on by contemporary naturalists when they discover new species today. Take, for example, the millions of creatures with jointed legs, from crabs and lobsters to beetles, bees and spiders. First, these are loosely classed together as *Arthropoda* (meaning jointed legs), then into orders of similar creatures (e.g. all the spiders belong to the order *Araneae*), then to family (all jumping spiders to the family *Satticadae*): finally they are pinpointed to a species – say the English jumping spider, sometimes known as the zebra spider, because of its striped back – *Saticus scenicus*.

continued

Opposite: *The splendours of the underwater world.*

NEVILLE COLEMAN
Underwater identification

continued

With a little practice and the help of a Latin dictionary you will soon start to spot the hidden clues incorporated in these Latin names: our plaice, for example, is a flat or plate-like fish *platessa*. Finally you sometimes find a Latinized personal name tagged on to the end of the species name – this tells you who found the organism; Linnaeus described so many new species that these bear only the initial 'L'.

deal of attention because of their extensive fossil record. Trilobite means 'three-lobed animal' and it resembled a centipede with three distinct parallel sections to its body. There were trilobites of every conceivable shape and form; some had horns, others had prickles. In time the three-lobed animal was superseded by other more complicated forms of trilobite; fossil trilobites have been found with up to 45 segments in their abdomens and tails.

The marine trilobites were the dominant form of life on Earth for almost 100 million years. Cartilaginous fishes, then bony fishes, fishes with gills, fishes with lungs, frogs, reptiles, mammal-like reptiles and mammals evolved by natural selection over the millenia, from these primitive marine creatures.

To appreciate just how much of our ancestry we owe to the sea, and how complex a biological laboratory it was and is, you need only consider the wealth of life which still exists on Earth that is connected, in one way or another, with water. In the five main taxonomic groups below note how many have an affinity with water:

1. MONERANS. The blue-green algae and the bacteria. These are the microscopic single-celled organisms found almost everywhere. There are about 4000 species, many of which live in water.

2. PROTISTA. The protozoans and the Phylum Protophyta. Single-celled and all but invisible, these are more commonly known as the planktons. The group contains about 50,000 species, all of which live in water.

3. FUNGI. The true fungi and the slime moulds are land organisms, although the strange slime moulds, which can 'crawl' about, need damp conditions. There are at least 100,000 species.

4. PLANTS. A huge group which begins with the algae – green, red and brown, numbering some 12,000 species with all but a handful of the most primitive green algae permanently resident in water. The other plants are organized into the mosses and liverworts, club mosses, horsetails, ferns, conifers and flowering plants. It is only possible to guess at the number of plants on Earth; new species are continually being discovered. A great many either live in or have an affinity with water.

5. ANIMALS. Consist of sponges, jellyfish and sea anemones, bryozoans, flat worms, nematodes and

rotifers, true worms, arthropods, molluscs, starfish and sea urchins, millipedes, centipedes, insects, spiders and scorpions, crustaceans, jawless fish, sharks and rays, bony fish, amphibians, reptiles, birds and mammals. Again, we have no accurate count of the numbers of animals on Earth, but a simple glance down these groups (or phyla) reveals a distinct leaning towards the sea for many of them.

There is general agreement that the most advanced animals in the oceans are those with backbones, be they fish or mammals. Less advanced creatures, like jellyfish or anemones, do not have a backbone. Sometimes advanced and primitive characteristics may occur in the same animal. For example sharks have some advanced characteristics but they lack some of the modern 'extras' like swim bladders.

Collecting and identifying animals and plants (see 'Linnaeus') is one of the most important tasks of nature-watchers. Our first sea-watcher, Neville Coleman, has undertaken a formidable project of this kind.

NEVILLE COLEMAN:
Cataloguing Australasian coastal life

Neville Coleman, Australia's leading underwater nature-watcher, is an amateur who has spent seven years in 'the biggest university of the world' – the beaches and shallow seas off the Australasian coasts. His purpose in life: to log the entire underwater flora and fauna of the Australasian coastal shelf.

Sandy beaches and shallow seas would not, at first glance, appear particularly fascinating, but Coleman knows better. 'It doesn't matter where you go on any seashore in the world, it's impossible to become bored,' he told us. 'You only have to keep your eyes open and be aware, and in a short time you will find lots of exciting things.'

The traditional way of cataloguing marine creatures is to collect specimens and store them in glass jars. Neville Coleman decided that an underwater photograph better preserved the true colours and forms. He started, single-handed, the *Australasian Marine Photographic Index*, which now contains more than 35,000 transparencies, all identified and cross-indexed. The Index is the product of more than 10,000 dives. Neville claims to be able to recognize more than 6000 species from memory.

Once his over-riding ambition in life was to find

Above: *Neville Coleman: impossible to be bored in the sea.*

Below: *Sydney, Australia.*

NEVILLE COLEMAN
Underwater life

Below: *Manta rays give the lie to the legends of sea monsters, even though the bigger ones can weigh four tons. Nervous and graceful, they feed only on tiny plankton.*

one new species. Today 10 (including Coleman's Urchin Shrimp, *Peniclimenes colemani*) have actually been named after him, because they were previously unknown to science.

Neville Coleman's message to nature-watchers could not be simpler: venture in with an open mind and you will discover ecstasy. This emotion is shared amongst those who explore the gardens of the sea, as we will demonstrate. However, first we should dispel a few myths that might be standing in your way.

The monsters of the deep
The more you get to know the sea, the more this popular label becomes a misnomer. Because the sea is such an alien environment for the landbound 'naked

ape' we have a fear of creatures that might lurk in the depths of the sea. If you leave out the mythical creatures such as the Loch Ness Monster, animals qualifying in size to be sea monsters include some whales, sharks, rays and a giant cephalopod, or squid.

The whales can be discounted immediately. Over the last decade the cetaceans, of which the whales are the largest members, have earned a reputation for gentleness in their relationships with the human race. The very largest whales do not even have teeth. Of the toothed whales (*Odontoceti*) only one, the killer whale, is reputed to be dangerous. Work done this decade with killer whales has shown, however, that these animals also behave with extreme tolerance and gentleness when in contact with humans. They are a popular attraction at most 'dolphinaria', where they may be seen engaging in rather idiotic circus tricks which would not be possible were the term 'killer' justified. They do hunt large food-prey in the wild, but by that definition every carnivorous animal on Earth is a killer. They are certainly not monsters.

Rays glide through the sea like huge bats, and the big ones such as *Manta mobula* grow to 20 feet in length and weigh up to four tons. They are also plankton-feeders; their common name 'devil fish' is emotive nonsense.

There are some very large squid in the sea and some even larger myths woven around them. They are the food of the largest toothed-whale, the sperm whale, and some 'monstrous' submarine battles must take place in the depths between these giants if the marks left on the sperm whales are due to the suckers of the giant squid. But none of the stories of huge tentacles reaching out of the sea to grab sailors (and even ships) has ever been substantiated. If the giant squid is indeed a monster, it is one that lives in another world.

Sharks have a poor reputation, but one which is ill-deserved, according to another eminent nature-watcher, Dr Eugenie Clark.

On the Beach
A beach is a graveyard of smashed rocks and the remains of hard-shelled sea creatures, including bivalves like cockles and mussels, the shells of snails and a variety of skeleton fragments – sea urchins, corals, bryozoans and other invertebrates.

The most distinctive feature of marine beaches compared to other marine habitats is their vegetation: they support may seaweeds and some 30 flowering plants which are often found in great underwater meadows. Off the coast of Denmark, for example, these meadows produce about 25 million tons of fresh plant material each year, which is more than that country's annual production of land-based forage.

Marine flowering plants flourish in warm tropical conditions – hence Neville Coleman's avid interest in them – but European divers can still see spectacular gardens close to their own coast. The *Zostera* (Eel Grass) pastures off the coast of Holland cover about 60 square miles. In the Mediterranean, other common flowering plants also form an almost unbroken underwater thicket round the coasts. Vast areas of the tropical Atlantic coast are covered underwater with turtle grass, and biologists have calculated that the meadows provide a home for as many as 30,000 animals per square yard.

EUGENIE CLARK
Sharks

Above: *Eugenie Clark (photographed on her sixty-second birthday!) remains in love with sharks.*

Above: *Sea of Cortez.*

Early exploration

Eugenie Clark's interest in the natural history of the sea began, as did that of all the great nature-watchers in this book, when she was a child. Her mother was Japanese and they lived alone in New York City, about as far from nature as is possible. On Saturdays Eugenie's mother had to work a half day and Eugenie was left to entertain herself at the public aquarium in Battery Park. These two hours a week 'pretending I was a fish' bred a life-long passion. She asked for an aquarium for her tenth birthday and this developed into an intense hobby for both mother and daughter.

She learned to swim and snorkel dive and loved 'the feeling of going through water in three dimensions – flying like a bird'. The passion persisted through her school years and into college, and she finally found a place at an institute of oceanography in California where she was invited to make a dive using a helmet. 'This was the only equipment they had back in the 1940's – a "hard hat" as it was called, with an air hose to the surface.

'I loved it! I was fascinated to be right down there as if in a Jules Verne story. Unfortunately I had a little accident: I didn't seem to be getting much air, but as this was my first dive I wasn't sure how it was supposed to feel and, with everything being so fascinating, I just kept going until I actually started to pass out. When I fell forward, cold water rushed into the helmet, so I took it off and swam to the surface. The rest of the party were alarmed because they knew my air connection had broken. Fortunately, as soon as I had recovered my teacher suggested I should go down again right away, or I might never want to dive again. The next dive went fine. Since then, of course, I've mostly used scuba equipment which allows you to be as free as a bird underwater.'

Eugenie has experienced almost every emergency a diver is likely to encounter. Very early on in her diving career, she joined a party attempting to reach the bottom of a freshwater spring, more than 195 feet deep. Few divers these days would attempt such a dive on air because of a condition called nitrogen narcosis or 'raptures of the deep'. This can be avoided by breathing a mixture of oxygen and helium, instead of one containing a lot of nitrogen, but this knowledge also postdates Eugenie Clark's apprentice dives.

'I was down at about 200 feet (in very murky water) when I felt as if someone had opened a window and that I was breathing very fresh air. I also felt I was about to give birth to a child, and heard the sound of people's voices. It came to me in a flash that this was the raptures of the deep which I had read about in one of Jacques Cousteau's early books, and that I'd better go up.'

On a subsequent occasion she was exploring some very deep caves for archaeological remains when narcosis struck again. This time she knew what was happening but, completely disorientated, she lost her way in the gloom and almost ran out of air. She had the sense to swim upwards and eventually rose above the narcosis threshold (about 145 feet for most people). By then she was breathing her reserve supply of air, which lasts a matter of minutes.

'I thought "Eugenie, this is it", but then I felt the cave passage beginning to widen. I came to the cave mouth and made it to the surface on what must have been my last lungful of air.'

All this could have been avoided if she had been diving with a companion and using a guide line. 'I wouldn't even think of doing such a dive now without them,' she admits.

To See the Sea

The simplest underwater viewing device is a glass-bottomed boat or bucket. If you decide to go further you must learn to snorkel, then to scuba-dive. We would encourage you to try both. Snorkelling offers a glimpse of the underwater world; 'scuba' (standing for *self-contained underwater breathing apparatus*) allows you to be a part of a world that is a garden more beautiful than mythical Babylon.

The snorkel diver needs only a mask, a snorkel tube and a pair of flippers. Be aware, however, that breathing through a snorkel must be learnt and masks and snorkels can sometimes fill up with water which can promote panic. We recommend that you buy good equipment from a specialist shop and take a snorkel-diver course with an organization such as the British Sub-Aqua Club.

At the very least you should learn to be happy in your mask and snorkel in shallow waters before venturing out of your depth. You will also feel increasing pain in your ears as you dive deeper than 8–9 ft. This is a natural warning, and can be cleared by pinching your nostrils and blowing until you feel a change of pressure, and then the discomfort goes. It is well worth investing in a good quality mask with flexible rubber around the nose which allows you to do this. Never over-ride this pressure warning. It is possible to dive deep enough with a snorkel to get the unpleasant experience of a 'bend' (nitrogen gas bubbles in the bone joints).

Scuba-diving requires a lengthy period of training but is worth every minute of it. There are now scuba clubs wherever the best dives are to be found, you will be able to hire equipment, and join these dives, if you carry a recognized qualification like those
continued

Left: *Gulp! A great white shark sharpens it's reputation for ferocity on photographer David Doubilet. Eugenie Clark took the photograph from an adjoining cage.*

EUGENIE CLARK
Sharks

continued
issued by the various national scuba or
sub-aqua organizations.

If you decide to become an underwater
nature-watcher, you will need quite a
lot of gear. The clubs can advise on
equipment, and much is available
second-hand. If you buy second-hand
air cylinders (diver jargon: 'bottles' or
'tanks') check that they have been
recently inspected and pressure-tested.
Scuba-diving is dangerous for the un-
trained and there are, unfortunately,
some diving schools that are more
anxious to take your money than give
you the essential training. Schools that
claim to be able to give you basic
training in an hour or so should be
avoided, as should the equipment sold
by schools or diving shops which do
not require you to show a qualification
certificate.

As a general rule, do not go underwater with an aqualung unless you have
been taught in the use of a weight belt,
clearing a flooded mask, sharing a tank
of air, signals, and the vital drills for
clearing pressure and coming up. Never
dive alone.

Do not be put off by these requirements. They are much the same as the
safety rules of the road, and learning to
dive properly is much less difficult than
learning to drive a car. You do not have
to be young or especially fit, provided
you have the training and choose your
dives sensibly. Do not be put off either
by the common worry about claustro-
phobia. Once you relax underwater the
feeling is one of immense space and
freedom, not of confinement.

But it is not our purpose, and least of all Eugenie
Clark's, to frighten you away from diving. Older and
wiser, she now uses the very best equipment and
always shares her dives with experienced companions.

'Get properly trained, stick to the rules and make
sure you are feeling comfortable underwater. That
way you'll be able to handle any emergency and can
have this wonderful feeling and a fascinating time
underwater.'

Befriending sharks

Her passion for sharks dates back to those early days in
the aquarium in New York. 'They didn't know how
to keep the water very clean. If you pressed your face
right up against the glass it felt as if you were right on
the bottom of the ocean, an ocean full of these superbly powerful creatures that moved around almost
effortlessly. I find sharks very beautiful, as do most
people who see them underwater: they are stream-
lined, they move with grace, and when you see them
in groups it is a really superb sight.'

Eugenie believes there is no more maligned or misunderstood creature, and that this ill-informed killer
image is being used to justify terrible attrition on the
sharks of the world. 'Sharks need protection from
people,' she snapped. 'If I had my way I'd have them
carry people-repellent rather than the other way
around.'

The killer image, she claims, is the work of the
ignorant or the exploitive. 'Sharks are timid, not
cowardly,' she emphasized. 'Real enthusiasts count
themselves lucky when they see a shark here and are
disappointed if they don't. It will only happen if you
are careful and considerate: keep really quiet – hold
your breath because they will even shy away from air
bubbles – and cling to the reef, making yourself part
of it.'

Eugenie further believes that sharks are considerably more intelligent than is suggested by their image
of being the vacuum cleaners of the sea. She taught
them to use signalling equipment as part of a long
series of colour and shape differentiation experiments.
'In fact, they learnt the difference between horizontal
and vertical stripes in the same number of trials as it
takes to teach a white rat to do the same thing, and
white rats are considered to be quite intelligent.'

Mention of the word 'white' provoked the obvious
question – were great white sharks also misunder-

stood? Eugenie had already shown us the picture on page 19. She was in the next cage and that same day a great white shark had attempted to bite her head off. It was her own fault, she insisted. 'We had deliberately set out to attract them with loads of chum (bloody shark bait).

'It is true that the natural prey of the great white shark is sea lions, which are about the same size as a person, and the shark might make a mistake. We also know that if a great white shark does bite a person, it seems to sense that it is not its food and spits it out.

'The trouble is that even if a great white shark takes a bite and then says "Oh, I don't want that!" you are already in a lot of trouble because its teeth are so big they have usually done considerable damage. For that reason it has to be regarded as the most dangerous predator in the sea, but it is not normally aggressive; in fact, it's quite shy.'

We insisted, nonetheless, that the moment she realized that she had left the top of her cage open and a frenzied great white shark was on its way in, must have been the most dramatic moment of her adventurous life.

EUGENIE CLARK
Sharks

Below: *All aboard the most thrilling ride on earth – Eugenie Clark astride a 60-foot whale shark. Flipped off the tail, she managed to obtain this purchase on the fin until her legs were skinned and the shark plunged to depths where no human diver dare venture.*

EUGENIE CLARK
Sharks

'No,' Eugenie smiled, for she is well used to fielding vicarious questions about sharks. 'My ultimate moment underwater was when I was able to ride a shark, 48 feet long, in the Sea of Cortez in California last year (when she was 61 years old!).'

This was a whale shark, by far the largest fish in the sea, but a completely harmless plankton-feeder. 'They come at you through the water like a space ship, with shoals of attendant fish riding on the bow wave, feeding and swimming in and around the gills, or tagging on for a free ride.'

Eugenie went out in a boat directed by a spotter aircraft and she and two photographers were dropped in front of their whale shark. Ever the most intrepid diver, Eugenie landed on the huge head – 'being careful not to slide into the mouth which could easily have accommodated four people' – then hung on to the huge dorsal fin – 'like a streamer'.

'My arm got so tired I had to pull myself up and straddle the fin, riding it like a horse. I wasn't wearing wetsuit pants and after I'd been there for a while I noticed there was blood coming from the inside of my knees, the shark's skin was so rough. The only possible position for a ride was a spot at the base of the tail but when I got down there it just flipped me off. By then my scuba tank had fallen down, I lost a flipper and my mask came off: I was in quite a mess, but the whole experience was just unbelievably thrilling.'

So she took the next possible opportunity to do it again. This time the whale shark did not stay on the surface but started to head down into the depths, leaving Eugenie's companions and her surface support-craft far behind.

'It was incredible. We went down and down and it started to get dark and my depth gauge told me I was approaching 170 feet. I started to feel the raptures of the deep coming on and thought "maybe it's time to go". Finally I had to let go and swim up, but that was the most thrilling dive I ever had.'

So what is left? If we stick with the definition of a monster as a creature to be feared by humans, there are a few small monsters in the sea. These creatures are probably responsible for most of our fear of marine animals, even though the majority do no more than sting. They have been studied in some detail by Dr Struan Sutherland.

STRUAN SUTHERLAND
Small dangers in the sea

Dr Sutherland, who works at the Commonwealth Serum Laboratory in Melbourne, is fascinated by the whole range of creatures that fit the general definition of 'monster', and his studies have greatly increased our understanding of dangerous animals. His chilling revelations about the killing ability of many Australian land creatures are detailed elsewhere (see p. 104).

His laboratory is situated within walking distance of several popular bathing beaches and he took us 'for a paddle' in search of some of the marine creatures that humans really would do well to avoid. After poking around in a rock-pool for a few moments with a little shrimp net, he offered up a tiny squirming jelly only a few inches in diameter, marked with bright blue rings.

'Now isn't that a gorgeous little chap!' he announced with real admiration. We moved in for a closer look. He moved it away. 'Its bite is deadly'; he smiled.

The creature was a blue-ringed octopus, and if the ease with which he had collected his specimen was any indication, they were quite common, as Struan confirmed.

'The actual biting equipment is a tiny beak in the centre of its eight little arms. The food goes through the beak and upwards between the eyes – right through the brain, in fact. The stomach is more or less sitting on its head. Up near the stomach are two large salivary glands which produce a toxin. This runs down two tubes, again through the brain, then out through the beak.'

We acknowledged that a blue-ringed octopus could seem somewhat monsterish to smaller sea creatures. Struan seemed happy with that reaction.

'They do have a strange hunting technique,' he admitted. 'Sometimes they will fall on their prey and inject their poison, but more commonly they swim down on, say, a crab and spray it with toxic saliva. The crab literally breathes in the poison while the octopus sits back and waits. After a few minutes the crab is paralysed and then the octopus moves in and pulls it apart. The octopus has a very delicate skin so it actually prefers to do this rather than engage in direct confrontation with a highly active crab.'

It is a poison of great sophistication which attacks the outer lining of the nerves, blocking the passage of vital neural messages which leads to death. (The

STRUAN SUTHERLAND
Dangerous sea creatures

Above: *Struan Sutherland: revealing Australia's most poisonous secrets.*

Below: *Melbourne, Australia.*

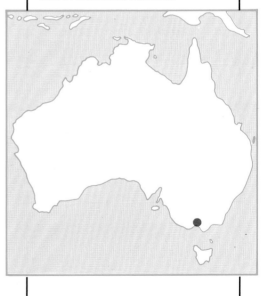

STRUAN SUTHERLAND
Dangerous sea creatures

Above: *This tiny blue-ringed octopus paralyses its natural prey (such as crabs) with a venom so deadly humans have also succumbed to it.*

main component of the toxin (tetrodotoxin) appears to be similar to that present in the flesh of many poisonous fish.) A mature blue-ringed octopus contains a large quantity of venom in its salivary glands, enough to cause the paralysis of some ten adult men. The bite itself is relatively painless but within minutes the victim notices a tingling sensation in his tongue and lips, and will soon have difficulty seeing or speaking.)

'It has a most peculiar effect,' Struan explained. 'We have interviewed victims who have been aware of everything that is going on around them, including the arrival of the ambulance. One of them even heard the driver say: "It looks like this poor chap has had it!"'

Needless to say, it was emergency treatments developed by the Serum Laboratory which allowed this particular victim to live to tell Struan the tale, even though there is no antidote for the octopus poison.

Artificial respiration, such as mouth-to-mouth resuscitation, must be given or the victim will become unconscious, and eventually die of heart failure due to lack of oxygen. If paralysis has occurred, adequate artificial ventilation may be required which sometimes has to be continued for some hours.

To set our minds at rest Struan reminded us that, apart from the blue-ringed octopuses, Melbourne beaches were 'pretty safe'. If we wanted to see some of the 'truly horrifying' Australian sea creatures we would need to visit the warmer northern coastline.

'Worst of all is the Box Jellyfish or Sea Wasp (*Chironex fleckeri*). It's the most dangerous jellyfish in the world; in fact, the only jellyfish to have killed anyone: it has killed more than 70 people this century. It wasn't identified until 1956. People would go swimming in our tropical waters, particularly on warm overcast days, suddenly let out a scream, run out of the water, and then collapse on the beach. They would be carrying wheal marks as if they'd been whipped and often they would die within a minute, or at the outside four to five minutes, of leaving the water. The stings were caused by a large jellyfish with a bell the size of a big bucket and tentacles stretching out several yards.'

There are two separate components to the toxin: one causes severe tissue damage and pain, while the other rapidly enters the bloodstream and causes respiratory and heart failure. It has been estimated that the adult box jellyfish contains enough venom to kill at least three men. If the victim survives but is not given antivenom, the part of the skin that has been stung ulcerates and often becomes permanently scarred. Struan's Institute has now isolated these toxins and produced an antidote. Because of the speed with which the toxins work, however, it is almost impossible to treat victims in time.

'We also discovered a terrible fallacy,' Struan revealed. 'It was once thought that alcohol neutralised the tentacles which adhered to the victim when they broke free of the jellyfish. In fact, if you pour whisky or vinegar over those severed tentacles they react by firing poison and the victim collapses again!'

Like the blue-ringed octopus, the very potent poisons employed by the box jellyfish have been evolved to protect it from the death throes of its prey. They live mostly on prawns which the jellyfish toxins immobilize in a second or two, allowing the food to be drawn up into the delicate bell where it is then digested.

When it comes down to it, Struan is rather proud of his 'monsters'. 'If you look at the box jellyfish objectively, it is incredibly beautiful and, for a jelly-fish, very advanced. It has eyes and can swim quite fast.' We sat on the rocks and watched blue waves breaking gently onto the sandy beach. Struan's mind was still focused beneath the waves.

'There are 26 species of poisonous sea snakes in Australian waters,' he said with real pride. 'In fact, all the most important and most dangerous of the world's sea snakes are represented around this coast.'

But by comparison with other things found in this idyllic stretch of ocean, sea snakes were not to be regarded as a serious danger. 'We have quite a lot of fish that are highly dangerous,' Struan went on quickly, concerned perhaps that our interest might flag. 'I'm very fond of one called the "Nohu" or "waiting fish". It has 13 venomous spines on its back. If you stand on one of these spines it can penetrate right through the sole of the foot and produce very severe pain. We've heard accounts of natives attempting to drown themselves in an attempt to get relief from the pain . . .'

And so we left him to muse on the ingenuity of nature while we went in search of others whose fascination with the sea has made them realize that there are no monsters – no mindless killers – apart perhaps from Man himself.

The Lie-and-Wait Fish
The Scorpionfish (*Scorpaenidae*) rarely move, and have evolved camouflage colours suited to their role as 'lie-and-wait' predators. A subfamily, the Stone-fish (*Synanceia*), share these character-istics.

Lionfish (*Dendrochirus*) and Zebra-fish (*Pterois*) swim or drift slowly across the bottom and have more exotic colouring, particularly the lionfish which are the most beautiful of the coral fishes.

All these fish have venomous dorsal, anal and pelvic spines, fed by poisons stored in tissues or glands alongside or at the base of the spines.

The pain experienced from a spine wound from some of the stonefish and

Above: *No-nonsense names are a feature of Australian fauna, hence* Synanceia horrida – *the horrid stonefish – whose spines carry painful venom.*

lionfish varies in degree, from bee-sting intensity to unbelievable agony.

Victims in severe pain should be taken to hospital after receiving first aid: it would help to immerse the limb in hot water.

2. CORAL-WATCH

Tropical islands evoke romantic images for all of us, but if you mention 'island ecosystems' to marine nature-watchers worth their salt, their eyes light up. They know that these islands can be primitive pockets of nature cut off from outside influences and alien contamination, like little Shangri-las.

Island Shangri-las have four main ingredients: a balmy climate, sparkling blue seas, shining white beaches and a coral reef; these essentials are only found together between latitudes 30°N and 27°S.

The coral structures to be found in this zone are the most extraordinary creations on Earth. They dwarf in size, complexity and variety anything built by Man. Thousands of islands are composed entirely of coral: the largest coral formation on earth, the Great Barrier Reef off the east coast of Australia, is more than 1500 miles long. This mass of living 'rock' has been created by millions of tiny animals, called coral polyps, secreting limestone around their bodies for generation after generation.

The first coral reefs were laid down in the seas of the Triassic period 200 million years ago; they are now buried beneath central Europe, graphic evidence of the upheavals and subsidences that the Earth has experienced.

Charles Darwin and an American, J. C. Dana, were the first people to propose a correct theory for how the main coral structures (fringing reefs, barrier reefs and atolls) are formed, in *Coral Reefs* (Darwin, 1842). They argued that fringing reefs are the first structures to be formed in the shallow waters around islands. These become barrier reefs as the landmass starts to sink, and when the island has sunk completely a lagoon surrounded by a ring of coral reef is left, called an atoll. Dana's contribution to the theory was to provide evidence that the level of the sea had altered around many Pacific atolls. More recent research has indicated that the Darwin-Dana theory may be too complex and that the reefs are simply necklaces of coral hung round emerging volcanic rocks; the extent and type of growth of the coral structures is controlled by water conditions.

Everyone agrees that the corals are the 'jewels of the sea' with radiant colours in settings and designs

Opposite: *Beautiful coral gardens ablaze with colour.*

of every conceivable shape, from giant fans to huge brain-shaped orbs. Some corals are soft and some reef structures are not corals at all but other animals that have evolved limey, hardened stems and rival the true corals in the complexity of shape and colour. Even the separation of flora from fauna has become blurred in this confusion of form and colour. Anemones (which are animals) look more like plants and are almost indistinguishable from some of the soft corals. There are the flower-like heads of bristle-worms, and shell-less molluscs and sea slugs whose naked gills wave like the petals of exotic orchids.

The behaviour of some of these creatures (especially the vermiform invertebrates) is equally exotic. The female of the echiurid worm, *Bonellia viridis*, is three feet long, and in a cavity within her live the inch-long males of the species. These males enter the female's body at the larval stage before their sex is determined. Any larva coming into contact with a female becomes male by the action of a hormone secreted by the female. Those larvae not coming into contact with adult females develop into females themselves.

Such is the multiplicity of form that we are discovering new types and species all the time, and at the moment it is only possible to draw up a very general list of the flora and fauna on tropical reefs. These include coral polyps, calcareous algae, countless species of fish, and a host of invertebrate animals that

Right: *In the reef gardens of tropical oceans, the dividing line between plants and animals becomes very blurred. This is, in fact, a giant tube worm.*

either crawl, run, swim over, bore into, stick to the reef or remain sessile (permanently fixed to one spot).

Unfortunately we humans only began to recognize the fragility of these island ecosystems when it was already too late for most of them. The survival of the coral reefs off the islands of Hawaii in the Pacific Ocean was once in jeopardy but now, as a result of the conservation campaigns by people like Jack Randall, the future looks much brighter.

JACK RANDALL:
Underwater conservation in Hawaii

Hawaii is one of the great coral areas of the world. It is also the fiftieth state of the United States of America. Joining this union has caused radical habitat changes, particularly on the main islands. Sections of coastline have been given over to military harbours and concrete roads; ribbons of hotels have been built to accommodate American tourists, and plastic grass is rapidly replacing the real vegetation.

On the other hand, Hawaii is one of the best examples of an ecosystem that has started to fight back. Americans have become conservation-conscious in the last decade and the marine activists of Hawaii have been in the forefront of this activity.

Jack Randall is one of the founding fathers of the Hawaiian conservation rescue. He sailed to the islands from the mainland in a small ketch in 1950 and lived there, aboard his boat, for five years while he acquired an advanced degree in marine biology. He and his wife and their two-year-old daughter then set off on a fish-collecting voyage through all the little islands we remember from adventure stories of the South Seas: Raratonga, Tahiti, the Marquesas Islands and the Solomon Islands. Eventually he returned to Miami in Florida to take up a teaching position at Florida State University; later he spent three years in the Virgin Islands conducting marine biological surveys. He came back to Hawaii to run the Oceanic Institute; he then became Curator of the Bishop Museum, on Oahu.

Jack is an ethologist specializing in taxonomy (the principles of the classification of living things) of marine creatures, mainly fish. The Bishop Museum has 27,000 fish specimens in bottles, the stock-in-trade of a taxonomist. Many of these are 'holotypes' (first-finds) and 150 of them have been discovered by Jack Randall himself and share his name. (It is customary

Above: *Jack Randall: proud owner of a unique collection of tropical fish, many of which he was the first to find. He lives and works in Hawaii.*

Below: *Oahu, Hawaii.*

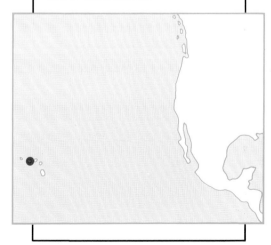

JACK RANDALL
Underwater conservation

for the scientific name of a species to include the name of the person that first discovered it.) He is widely regarded as the world expert on fish of temperate seas and he is the author of several books on fishes.

When we met Jack Randall in Hawaii, the first visit he proposed was to a busy beach on Oahu, the main island of Hawaii, called the Hanauma Bay Marine Reserve. Here we saw the effect of declaring the place a fish sanctuary. The fish are now quite tame but Jack assured us it was very different 25 years ago. 'If you saw a fish at all in those days it was only its tail heading away from you as fast as it could go.

'We started to see a dramatic change in the behaviour of the fish within a year or so of the place being declared a marine reserve, not just in how close they would approach but also in their numbers. They are now utterly fearless; it's marvellous, but that's what you get if you protect the fish and the habitat. No-one may harm the fish, collect corals and shells, or even take one grain of sand.'

On the beach Jack let us feed a moray eel to prove his point about the tameness of the creatures, but we could see his attention beginning to shift across the bay to the reef and the deep water beyond. 'You won't find any new species of fish here in shallow water,' he said. 'We collected and named all the shore fishes long ago and the total for the Hawaiian islands is now nearly 700, of which 300–400 may be called reef or shore fishes. But if you want to dive down into some deeper water . . .' We were delighted to take up his invitation.

Reef fish curiosities

Cleaner wrasse
Jack Randall showed us a group of fish called cleaner wrasse (*Labroides*) which are an example of one of the sea's most remarkable symbiotic relationships. Jack explained: 'The wrasse which live amongst the coral remove parasites and feed on the mucus covering a great variety of reef and shore fish. The latter seem to know where these cleaning stations are, and they actually jockey for position to get cleaned.'

Jack has even seen 'customers' drop in from the huge semi-pelagic shoals that swim by, later rejoining their shoal after being cleaned. Huge manta rays go to the station and allow the tiny four-inch-long wrasse to scour their gills; even aggressive moray eels take advantage of the service.

When business is bad, the wrasse advertise. They do a little dance, spreading their tails and oscillating the back part of their bodies up and down as a signal to the surrounding fish.

Jack pointed out that one of these cleaner wrasse was something of a wolf in sheep's clothing; not a wrasse at all, but a sabre-toothed blenny which has evolved similar coloration to the wrasse to mimic its appearance so that it can predate their customers. 'They hang around the cleaning station, acting just like the wrasse and hoping that the other fish won't notice them. The blennies have huge canine teeth projecting from the lower jaw, and this particular species tears off and eats the scales of other fish which are deceived by its colouring.'

JACK RANDALL
Underwater conservation

Below: *'Cleaner' wrasse (one is at work here inside the gill slit) run 'stations', at which large fish queue up for attention.*

JACK RANDALL
Underwater conservation

Above: *Parrot fish appear to eat coral;
in fact they refine out the algal nutrient
present after grinding the coral to sand
in the rear of their mouths.*

The goby fish–snapping shrimp partnership
Subterfuge and symbiotic relationships are being practised all over the reefs by living organisms. On one of his voyages, Jack Randall observed a symbiosis between a goby fish and a snapping shrimp that was remarkable in its complexity. The goby appeared to be using the protection of the shrimp's burrow which the shrimp spent its life clearing.

'When we first saw this we assumed that the goby was a "free-loader", merely letting the poor little shrimp work away like a bulldozer all day long. Now we know that the goby plays an important role as sentinel and alarm signaller for the shrimp. The goby has very keen vision and its lateral line sensing system can pick up the low frequency vibrations made by an approaching predator. When danger comes along, the goby backs its tail into the hole to prevent the shrimp from coming out and flutters its tail as a signal. If the danger is great, the goby spins round and dives head first down the hole, pushing the little shrimp in with it. We don't think the shrimp can see very well, because it won't come out unless it can touch the goby, and while out it maintains frequent tactile contact with its antennae. So both individuals benefit from this partnership.'

Parrot fish
Moving on we encountered a brilliantly coloured group of fish which appeared to be eating the coral. These were parrot fish, an appropriate name for creatures with a beak like a parrot and colours to match.

Charles Darwin saw them at work when he was sailing in the *Beagle* and he assumed that the parrot fish ate coral. We know now that this is not quite the case. A couple of very large species have been observed feeding on live coral, but most parrot fish scrape algae from dead coral for food, using a unique system.

The algae are notoriously difficult to digest. The parrot fish scrape off the coral (using their fused teeth or beak) then pass it back to the rear of the mouth into the equivalent of a bird's gizzard, a pharyngeal mill, where the rock and algae are ground to a powder. Having overcome their processing problem, the parrot fish are left with a very awkward waste product: coral is almost pure calcium carbonate which, if mixed with stomach acids, would form carbon dioxide gas and blow the fish up. The parrot fish have eliminated this problem by dispensing with a stomach: coral gravel passes straight from the mouth into the intestine where the algal nutrient is extracted and the gravel passed on through the body and excreted. It may seem a difficult way of getting a meal but in the heavy competition of the reef it has given the parrot fish a unique feeding niche and they are doing very well as a result.

Parrot fish leave little grooves in the corals from which they have taken algae, making it easy to spot where they have been feeding; but as far as anyone knows, they do no real damage to the reef.

Parrot fish belong to a family of fish (including wrasse and groupers) in which individuals have the remarkable ability to change from female to male. Jack, who has done a great deal of work on the phenomenon, explained. 'They all start out as females. In the case of the groupers they reach a certain age and simply switch over from being female to male. Wrasse and parrot fish start out in what we call a ''drab'' stage in which most of them are female. When they reach a certain size and age (Jack believes age is the important factor) they change sex and often take on a much brighter body colour. Green is the dominant colour and in this outfit they are males.'

Once male, the fish establish a territory and build up a harem of females. Work done by Australian scientists with cleaner wrasse on the Great Barrier Reef has shown that if the male is killed, the dominant female starts to act like a male, then changes to a male and takes over the harem.

This change is the result of hormones and nobody knew exactly how long the process took, nor did they know that such a strange metamorphosis could be occurring, until fish were found in the intermediate stage. 'You don't see this very often because in the fish's 20-year life span it will spend only two weeks in the intermediate stage,' Jack explained.

This is just one of the cases in which scuba-diving has advanced scientific knowledge by making detailed observations of the behaviour of sea creatures possible.

The crown of thorns starfish

One of the most serious threats to the survival of coral reefs comes, not from Man, but from the crown of thorns starfish. It is covered with poisonous spines – hence its name. Under normal circumstances, the number of coral polyps it consumes would make no

JACK RANDALL
Underwater conservation

Below: *Crown of thorns starfish do destroy coral polyps; the corals kept ahead of the starfish until human pollutants started killing off starfish predators. Now massive coral formations like the Great Barrier Reef could be threatened by crown of thorns.*

difference to so vast a conglomerate. However, crown of thorns starfish have been experiencing bizarre population explosions in recent years, which have caused considerable alarm among environmentalists. Thousands of these starfish have been known to suddenly descend on a reef, given a favourable current. They cover the coral and extrude their stomachs to digest the polyps directly: in their wake they leave a dead coral reef.

A decade ago the people of Australia became gripped by panic when the deadly cloud of crown of thorns was found on several parts of the Great

Barrier Reef, and they had good cause. The climate of the eastern half of the country (and arguably of the whole country) is conditioned by the existence of the huge reef. If it were to vanish, the climate, indeed the east coast itself, would alter.

Jack Randall, who has been studying crown of thorn infestations on island coral reefs for more than a decade, was one of the senior experts called in and was among the first to suggest that, in a sense, the plague was our own fault.

'We saw our first crown of thorns explosion on the reefs off the island of Guam in the Pacific Ocean in 1968,' Jack said. 'About 95 per cent of the corals of the entire western coast of Guam were destroyed on that occasion. They tried to control the starfish with divers, but there just weren't enough divers to manage.

'The big question has always been, what causes these enormous population explosions? For a long time it was simply assumed that this was a natural cycle of abundance that has been going on since the beginning of time.'

Suspecting that this was not the case, Jack made a study of the starfish populations of 28 South Sea islands, and he came to the following conclusion: 'The islands with the largest populations, the most developed agriculture and the use of pesticides (in his study they were Tahiti and Raratonga) also had the largest number of these starfish. By contrast on those islands which were uninhabited, or that had few people, we found very few of these starfish.' Randall found confirmatory evidence for his theory when he was asked to look into the Great Barrier Reef infestations. 'The first major infestations all occurred near centres of population where there was very heavy agriculture and where the Reef came closest to Australia.'

Jack Randall believes that pollutants from fields dressed with fertilizers and pesticides, and waste from human population centres are to blame. 'Chemical pollutants like the chlorinated hydrocarbons are almost certainly harming the predators of these starfish – most likely the fish that eat the young starfish under ''natural'' conditions and keep the starfish population in check. All of a sudden huge numbers of juvenile crown of thorns survive and develop into adults.' Nature, he feels, would not allow this to happen even in very occasional periods of food abundance, given the terrible damage the starfish plagues do to the coral habitat of so many other creatures. Nature

JACK RANDALL
Underwater conservation

keeps things in equilibrium. So there must be external, man-made, disturbances somewhere.

'There probably are cycles of abundance, but not to the degree we saw in the attack on Guam. Nature doesn't allow one animal in a complex community like the coral reefs to get so widely out of balance with the rest: there are always repercussions.'

Changing human attitudes to the sea

Jack Randall's concern has become increasingly focused on the casual way we treat the sea and its inhabitants. Marine pollution is a major concern in Hawaii and may already be affecting one of the most commercially important fish on the islands, the tuna. Such creatures should be protected internationally, he believes, as a world food resource.

Hawaii has imposed strict controls on imports of marine stocks. 'We saw a number of introductions backfire,' Jack observed. 'They brought in the Marquesas sardine for tuna bait but a few other fish came along with the sample, including we think, one little mullet. Five years later it was quite abundant: now it is competing, and might even be replacing, a large, commercially important, mullet.' The blue-striped snapper was also introduced from Tahiti and the Marquesas, probably for the aquarium trade, and is now very common everywhere, including the marine reserves. 'I think it's undergoing a population explosion too,' Jack said ominously.

He sits on an advisory board of the Department of Agriculture for Hawaii, which has the power to veto animal or plant imports that could be potentially dangerous to the native habitat. 'It may be something as innocent as oysters from the USA,' he pointed out. 'But if there is any risk of man putting something into the sea that would disturb our own natural environment or the organisms living therein, I will try and prevent it. We should not tamper with nature.'

Jack Randall is one of many concerned nature-watchers on Hawaii; together they have mounted a tough rearguard action to ensure no further damage is done to these beautiful islands which seems to be working, at least below the waves.

Unfortunately, Hawaii is the exception to the international rule. The majority of the most wonderful coral gardens in the world are tucked away in forgotten corners where conservation has low priority.

EUGENIE CLARK:
Exploring the Babylon of the sea

EUGENIE CLARK
Diving in the Red Sea

If you were to come across anything remotely resembling a coral garden on land it would seem mystical: a freak show of bizarre plants amidst a collection of surreal statues, all daubed in fairground colours.

And of all of these, Ras Mohammad is the most beautiful. It is a rocky headland, overlooking the Red Sea, at the southern tip of the Sinai desert. The desert has an awesome beauty but is seemingly devoid of life and therefore one of the last places where you would expect to find a unique wildlife habitat, or where anyone would think to provide a wildlife sanctuary. Yet this wonderland can be seen from the Ras Mohammad headland. With the empty desert behind you, and the scorched, barren rock beneath your feet, you can look down into the glass of the sea at multicoloured corals and a wealth of magnificent fish as if they were swimming in an aquarium.

Or, as we did with Eugenie Clark on her sixty-second birthday, you can dive there and experience possibly the most colourful nature-watch on Earth. For her it is, quite simply, 'the best dive in the world'.

Eugenie Clark has been making pilgrimages to the Red Sea for 25 years. She made her first visit to an Israeli research centre 25 years ago as a Fullbright Scholar, learning then why this particular stretch of water was so unique.

'The Red Sea is long and narrow,' she explained. 'It has a very narrow opening into the Mediterranean through the Suez Canal and another opening into the Indian Ocean at the southern end. There are no permanent rivers running into it and it is part of a desert ecosystem. In effect it is a great evaporation basin: the saltiest expanse of water on Earth. There is a crack, or trench, along the sea bottom which turns at Ras Mohammad, separating the African and the Asian continental plates. This trench is volcanically active and produces hot brine. That again makes the Red Sea unlike any other sea because it has no deep cold water.

'Immediately alongside the coastal reefs, however, is very deep water going down hundreds of feet. Any sediment sinks into these depths, leaving the surface water marvellously clear. A hundred feet is considered poor visibility in this area and when it's good it is well over double that.' (Most British divers consider 50 feet of visibility 'good'.)

Above: The Nature Watch team, equipped for underwater, preparing to explore Ras Mohammed, in the Red Sea, the most exotic marine garden in the world.

Below: Ras Mohammed, Red Sea.

EUGENIE CLARK
Diving in the Red Sea

'Here the most beautiful and interesting dives in the world can be made. You can go to the Great Barrier Reef and see 940 miles of gorgeous coral but everything is very shallow and spread out. Here at Ras Mohammad, everything is concentrated: you can go on one dive and see perhaps a thousand different living creatures all in one spot.'

Eugenie has dived in every tropical ocean in the world but remains staunchly in defence of Ras Mohammad as the dive site *par excellence*. 'Just taking the number of corals, you won't find such a concentration in other places, and Ras Mohammad has a tremendous number and variety of fish too. The very exotic parrot fish sometimes shoal here in their thousands. The crushed coral they pass through their bodies (see p. 32) forms little beaches that provide habitats for other fascinating creatures, such as garden eels.

'We're finding new species all the time. I came across one whose generic family I knew, but whose presence had never before been recorded in the Red Sea. It buried itself in the sand but I managed to retrieve it and bring it to the beach where my little son Nicki was playing. When it was confirmed as a new species I named it after him.'

Diving with Eugenie Clark
By now we were itching to get into our wetsuits but Eugenie had a small chore to do before she could go down and visit some of the fish that occupy fixed territories and have become her friends.

'I need to string some eggs for George,' she said.

The crew running our diving boat knew what she meant. 'George hasn't been around for a while,' they said. 'But we think you'll see Georgina on guard by the wreck.' (We were tied up to the wreck of a freighter which came to grief on the edge of the reef a few hundred yards offshore from Ras Mohammad.)

Georgina is a giant blue-green Napoleon wrasse. Napoleon wrasse are the largest members of the wrasse family. Georgina is over three feet long and half as deep in the body with a lugubrious mouth and a lump on her head. Apart from that she is just a fish, nervous and greedy.

The wreck site is a popular dive for marine nature-watchers who now come here from all over the world, and the fish are so used to them they can be fed by hand with scraps of squid, buns and, in Georgina's case (given the size of her mouth) hard-boiled eggs.

Eugenie had decided this would be an ideal way of introducing us to the wonders of life underwater at Ras Mohammad.

We dived and were soon experiencing a bizarre underwater pageant.

The uncommon garden eel
Imagine lying on a white beach. The blue sky has come down to lie around you like a blanket. Multi-coloured birds with no wings go past. Suddenly out of the beach rise thin, snake-like wands with tiny heads and big ears, as if from a snake-charmer's basket, and they begin to weave and dance to a tune you cannot hear.

You are 50 feet underwater. The shock of your arrival has passed and the garden eels have lifted from their burrows to resume the vital business of feeding on passing plankton. There are hundreds of them, each a yard or so apart, as if planted by a gardener with an eye for order; it is a magical experience.

Below: *George and Georgina, giant Napoleon wrasse, grew so tame that divers at Ras Mohammed were able to feed them by hand. (To avoid getting her fingers nipped, Eugenie Clark hung hard-boiled eggs on string for Georgina.)*

EUGENIE CLARK
Diving in the Red Sea

'Here is one of the largest garden eel colonies in the world,' Eugenie explained. 'The species was first described in 1934 (before the days of scuba) by an American scientist, William Beebe, as a "garden of eels" and the name stuck. People didn't believe him at first because he was unable to collect any specimens. He tried grabbing them and digging them out, but it was no good because they can literally swim through the sand and escape.

'Garden eels were regarded as a figment of his (Beebe's) imagination until scuba tanks came into use, enabling people to remain underwater for lengthy periods. Divers all over the world soon started reporting colonies of them. Now about 14 species have been described, all living in sandy areas in tropical seas.'

Above: *Not sea snakes, but eels living in burrows into which they retire apace at the approach of anything threatening. Here, Eugenie Clark emerges from the unique underwater Red Sea 'hide' built to observe these shy and graceful creatures.*

Eugenie and her students have conducted an intensive scientific study of the eels over a period of several years which has totalled more than a thousand hours of underwater observation. They have established that the eels feed exclusively on plankton and live a life governed by a dogmatic fixation to their burrows. No garden eel has ever been seen to leave its burrow, yet they engage in territorial fighting, mating, and all the features of life required for survival.

'The big males have a territory that is like a hemisphere over their burrow opening. They defend it as far as they can stretch without taking their tails out of the sand. Smaller males have smaller territories because they cannot stretch so far. They even fight from this fixed position, with the larger males keeping the smaller ones outside their range.

'The only moves may occur during mating, if a male is attracted to a female in a burrow just outside his range. We took timelapse photographs of a male engaging in courtship display, rippling his fins and making all kinds of movements towards a nearby female. In the next timeframe, the same male appeared next to the female. It is just possible that the male may leave his burrow, but I think that he goes under the sand and comes up alongside the female. We made casts of the burrows which indicated this might be happening. Most of the burrows were just deep enough to accommodate only the length of the eel, but some of the burrows went down off a side track that could accommodate a move to the female.'

Even so, the move is never permanent. Eugenie speculates that the male always returns to his original territorial position after a mating move (which can last for several days) because their positions are dictated by the available food supply. 'During the period the male and female are together, the amount of food they get from passing plankton is obviously cut by half because it has to feed two mouths, not the usual one.'

A short underwater swim from the Garden of Eels is a famous spot on the reef known as 'Anemone City' for the profusion of its coral structures and the breathtaking array of colours.

Symbiosis amongst anemones

Sea anemones are animals, not plants like their land namesakes. They belong to the same taxonomic group as the jellyfish and corals (*Cnidaria*) and they appear rather plant-like with their 'flowering heads'. Anemone fish (or clown fish as they are sometimes called) live in symbiosis with a group of deadly poisonous anemones (stoichactids). The fish hide amongst the amenones' stinging tentacles unharmed because of a slime covering them. This coating originates from the

Diving in the Red Sea

anemone tentacles; it prevents the release of the tentacle toxins when they wave against the slime-coated fish.

'The clown fish looks almost as if it is cuddling the anemones in a very affectionate relationship,' Eugenie pointed out. 'In fact, they are harvesting the protective coating from the tentacles and if they don't keep doing this they lose their immunity to the anemone toxins. We've tested it experimentally: if you dip a piece of cotton wool in alcohol and wipe that coating from the fish it will swim right into the anemone and be killed and eaten by it.'

Above: *The sea abounds with 'you-scratch-my-back-and-I'll-scratch-yours' relationships, but few are as complex as the deal between anemones and anemone (or clown) fish: the fish lure prey into the tentacles of the anemone, while the anemone offers them protection from its stinging tentacles by coating them in a protective slime.*

There is a price to pay for this protection. The clown fish spend their life deliberately offering themselves as prey, or bait, to fish swimming near the anemones, luring them into the lethal web of tentacles and so providing food for the anemone.

The lionfish
Possibly one of the most beautiful fish on the coral reef is the lionfish (or scorpionfish). But its brilliant colours are flags of warning: it is also one of the most deadly reef creatures. 'It can kill a person,' Eugenie confirmed. 'They have 13 spines, at the base of each of which is a poison sac. If you get a number of these

needles in your stomach it could be fatal, while just a few in your hand will put you in hospital.'

Divers should always keep their distance and not annoy the lionfish – 'if you annoy them too much and get too close they will actually charge, lower their heads and jam these spines into you.' They also have the unfortunate habit of using a recumbent diver as a hide. 'When you're lying in the open sand studying something for a long period, the lionfish will swim across from a coral head and cuddle under you, trying to camouflage themselves so they can catch the little fish swimming by. When you've been lying there for half an hour or so, you sometimes find you have six of these lionfish cuddling up beside you. It's a situation that requires very careful movement because if you startle them, they'll jab you.'

We hasten to add that in her thousands of hours underwater at Ras Mohammad, Eugenie has yet to be harmed by a lionfish's spines. Diving one night, however, she thought her luck had finally run out. 'We

Below: *Lionfish have thirteen venomous spines lurking beneath one of the sea's most elegant displays of* haute couture*; moreover they like to cuddle divers.*

EUGENIE CLARK
Diving in the Red Sea

Below: The Spanish dancer – the story of the ugly duckling in reverse – for in fact this flamboyant gymnast of the sea is a sea slug, admittedly with its gills worn on the outside. These nudibranch molluscs are best seen at night when they form a vivid focal point to the kaleidoscope of colour that is the tropical seascape.

were diving with the flashlight fish – a beautiful sight to see at night. They were signalling to each other and I was trying to study these signals by showing one of the fish its own image in a mirror. It changed its flashing pattern and became so intrigued by the mirror I was actually able to back off and lure it away from its group. Then I felt spines go into my backside and thought ''Up, and off to the hospital, fast.'' But I was lucky – it turned out to be a sea urchin.'

Night diving
The appeal of night diving is that the colours of the fish and coral are at their most vivid under strong artificial light. Nowhere else in nature is colour being used with such controlled abandon, so tastefully and yet so garishly, so stylishly and yet so formally, as on a coral reef at night. The behaviour of fish is also different at night: there are those attracted by the light, which come crashing into the pall of illumination like glowing comets from the darkness of space. Others like the blue, green and red peacock wrasse

are found asleep in the coral heads and can sometimes be picked up.

If you are lucky you will see a Spanish dancer, in Eugenie's opinion one of the loveliest creatures you can come across on a night dive. This animal is actually a slug; when our torches eventually located one of these extraordinary creatures pulsating down the side of a back-lit sea fan, we realized that Eugenie's extravagant description was no exaggeration.

Spanish dancers are nudibranch molluscs (sea slugs with naked gills). These gills are the pink flower-like structures on top of the creature on page 44. 'The locals call them "Bedir" after the most famous Egyptian belly dancer of all time. Sometimes you find little shrimps riding the dancers, as if they were taking a trip on a magic carpet.'

What can possibly be left for Eugenie, we wondered? She appears quite capable of diving for another 20 years. Therein lies her present incentive: to ensure that others will be able to share her enjoyment in the future. For almost a decade she has been urging the Egyptian authorities to declare Ras Mohammad a marine sanctuary, to be totally protected from fishing and redevelopment. This has been no mean task, since there are no nature reserves in Egypt and therefore no administrative structure under which Ras Mohammad could be declared a sanctuary and policed.

In 1981 Eugenie obtained a personal interview with the late President Sadat. She lent the weight of her international experience to ginger groups like the British-based Friends of the Red Sea, and interested affluent contacts in the USA. Then in 1983 the seemingly impossible was achieved.

'After a debate in the Egyptian Parliament which went on for about 18 months, they created a whole national parks system virtually in order to make Ras Mohammad a sanctuary, which it became by prime ministerial decree in November 1983.'

It is a splendid culmination to a life-long commitment to the sea, but Eugenie sees it as more a matter of paying off a debt to some very old friends. 'We get such wonderful enjoyment from the creatures in Ras Mohammad, the least we can do is to respect the habitat which, after all, is their home. We are guests in the sea.'

3. WHALE-WATCH

No one pretends that we are even close to the threshold of understanding the minds of whales, least of all Dr Roger Payne who, with his wife Katharine, has studied the 'singing' humpback whales and has made a crusade of whale conservation.

ROGER PAYNE:
In praise of whales

Like all distinguished nature-watchers, Roger Payne has been interested in nature for as long as he can remember. He told us that the great love of his youth had been owls. Later his schoolboy hobby developed into academic inquiry and his doctoral thesis was based on an elegant study which revealed, amongst other things, that owls could locate food prey from sound clues, in pitch darkness, to an accuracy of 1°.

In search of pastures new, Payne turned naturally to creatures that spend most of their lives in the darkest abysses of the earth, the deep-sea whales. In 1967, Roger and his wife Katharine went to Bermuda to record the vocalizations of Humpback Whales (*Megaptera novaeangliae*). Each year these whales leave the icy waters of the north for their calving grounds in Puerto Rico, passing Bermuda en route.

The 'songs' of the humpback whales

What exactly did Roger and Katharine record from the ocean as they drifted about in their sailing boat, trawling two hydrophones along the line of 'sound trails' that had been previously discovered by Roger and a sound specialist, Scott McVay?

No less than: 'All that night being borne along by lovely, dancing, yodelling cries, sailing on a sea of unearthly music . . . a vast and joyous chorus of sounds that poured up out of the ocean and over-flowed its rim. The spaces and vaults of the oceans, like a festive palace hall, reverberated and thundered with the cries of whales.'

Payne listened to tapes of humpbacks made by a Bermudan friend, Frank Watlington. When they analysed the tapes, Payne and McVay found that the long, complex patterns of sounds made by the whales were being repeated every 10–15 minutes – the

Above: *Roger Payne: 'I would not wish to live in a world without whales.'*

Below: *Hawaii.*

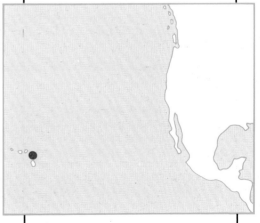

Opposite: *A humpback whale erupting out of the sea.*

ROGER PAYNE
Whales

Whaling

Records of true commercial whaling exist from the twelfth century, when Basques killed whales from open boats in the Bay of Biscay. By the sixteenth century their boats were operating as far afield as Greenland.

The British, the Dutch and the North Americans began whaling a century later. For a time the primitive nature of this assault, involving hand-thrown harpoons from open boats, kept the attrition within reasonable bounds, but this *status quo* changed abruptly with the mechanization of whaling, bringing improved ships and harpoon guns.

In the late 1700s the industry collapsed through overkilling in the North Atlantic and, 100 years later, the North Pacific was similarly denuded of viable stocks. By the beginning of this century, whaling in the higher latitudes for right and sperm whales had also ceased to be viable and, but for the invention of the explosive harpoon gun in 1868, and faster steam-driven ships (which enabled the rorquals to be pursued into the Antarctic feeding grounds), the industry might have collapsed entirely.

Slaughter was also speeded up by the use of factory ships and there is a record of 46,000 whales being taken in one season in the mid-1930s. Whales like the blue whale, which had earlier become very rare, suffered renewed attrition as their numbers increased and by the 1950s, were being caught at the rate of some 20,000 a year until they again went into decline.

The International Whaling Comission (I.W.C.) was set up in 1946 in an attempt to regulate a free-for-all that was quite obviously pushing the whales into extinction. But only in the last decade, in the face of growing pressure from conservationists, has the Commission agreed to real limitations on the whaling activities of those countries still in the business, primarily Russia and Japan.

continued

whales were 'singing'! These are 'songs' in the same sense as animals which make repetitive vocalizations (such as birds, crickets and frogs) are said to be singing. Humpback songs were immediately recognized as the most elaborate non-human vocal displays yet to be discovered.

Characteristics of humpback 'songs'

The whales sing continuously, without pauses between songs, sometimes for hours on end. A single song may span $5\frac{1}{2}$ octaves and include a vast array of sounds, from rapid trills to notes lasting up to 18 seconds, and it may have many different tonal qualities, rhythms and pitch configurations. Anything up to nine distinct 'themes' are sung in an unvarying sequence. Each theme consists of several identical or slowly changing 'phrases' composed of two or more sounds. Although themes are sometimes omitted, those that are used always appear in the same order. The pace of singing and the number of phrases and themes per song vary from whale to whale, and even between songs by the same whale. Songs thus vary in length, with the average lasting about ten minutes; songs lasting 30 minutes have been recorded.

What does all this tell us? A few things certainly that no-one had ever credited to animal intelligence hitherto. For example Katherine Payne discovered that these songs change constantly, a unique feature in animal songs. In any one area all the whales sing the same song. But as they sing they improvise, changes start to appear and old material is dropped, until after five to eight seasons, it has become an entirely new song. All the singers in one area must be revizing their songs more or less simultaneously to keep up with the current version. They probably do this by listening to each other and imitating their neighbours' song. All the singers in one area must be revising force in whale society, so a study of the songs might reveal aspects of their social structure and, through research on how they manipulate and change the songs, might also reveal information about the mental processes of the humpbacks.

After the initial series of recordings in Bermuda, the Paynes moved to Hawaii in 1975, where the humpbacks come closer to the coast, the water is clearer and calmer, and the season is not interrupted by a migration. Bermuda is not believed to be a breeding ground, but only a migratory way station for humpbacks swimming between their North Atlantic breeding grounds and their southern breeding grounds in the Antilles.

During the first season of their Hawaiian study they discovered once more that the songs became increasingly stable. Improvisations, new material and so on are first tried and tested by a remarkably democratic process, then moulded into the end-of-term piece by constant practice. When the whales first arrived in December, 83 per cent of the songs were missing one phrase and 33 per cent lacked two phrases. But by April all but 12 per cent of the whales were singing complete songs. They had also stopped producing incomplete or 'hybrid' phrases which, at the start of the season, were taking up 13 per cent of their singing, but by the end featured in only two per cent of their songs.

Apart from stabilizing through the year, the Hawaiian songs (like the Bermudan ones) grew in length. The Paynes' careful measurements and their dissection of the songs into phrases showed that the songs were lengthened by increasing the average length of each phrase, the number of notes in a phrase and the average number of phrases in a theme. By the end of the season a song which five months previously had lasted a little over seven minutes had been stretched to 16 minutes.

'This is particularly remarkable when you realize that the whale is breathing only once per song,' Roger observed. 'This implies that the change in the song may also be accompanied by a physiological change.'

While the whales of Hawaii were singing different songs to those in Bermuda, the rules for forming a song were the same. This implies that a knowledge of the laws of song-making is inherited by the whales, though whether the inheritance is genetic or cultural has yet to be established.

'Songs' are composed of two parts
Once they had collected a substantial body of information, the Paynes were in a position to study some of the nuances of humpback sound production. These suggested that the songs were 'informational', and also dual-purpose.

The whales sing more at night and the songs are composed of two frequency patterns: high-pitched high frequency sounds which are soft and varied, and low frequency expressions which are less varied and much louder (between 50 and 10,000 hertz). Payne

continued
Activists in the United States brought about the first effective limitation when the US Marine Mammals Protection Act was passed in Congress in 1972, prohibiting the taking and importing of marine mammals and their products (other than by a few Eskimos and Indians).

In the same year, a United Nations conference called for a ten year halt to all whaling – but this was rejected by the I.W.C., giving rise to accusations that the Commission had become a cartel for whaling interests. A campaign, led by Sir Peter Scott, was then launched to alter the I. W. C. position, and in 1982 success came when a motion was forced through in favour of a moratorium on all whaling from 1986 onwards.

Unfortunately I. W. C. members are not bound by the 1982 edicts if they lodge suitable objection within a prescribed period. The Governments of Japan, Norway and the USSR all lodged complaints and they are therefore not bound to adhere to the agreement. The fight to save the whales is still far from being won and we commend all nature-watchers to support this cause.

ROGER PAYNE
Whales

Right: *Whales, like this finback, are traditional nature-watchers, rising to keep an eye on humans and their boats when they have the time and the inclination. Unfortunately the historical response was most often a harpoon.*

turned to another of his research programmes with Finback Whales (*Balaena physalus*) – long-distance callers with songs only two or three monotonous calls long, very unlike humpbacks – for clues to explain the use of two frequencies. He concluded that they were being used for different transmission requirements.

The finbacks use low frequency calls of 20 hertz. Naval experts engaged in acoustic research initially refused to believe that an animal could make such a pure tone. This single blast of sound is followed by a set period of silence, lasting 12 or 15 seconds alternately. The special feature of this call is the great distance it can travel before being lost in the background noise of the ocean. Because of its simple structure, it has been assumed to be an identification signal, such as: 'There is a finback in this stretch of ocean.'

Even though they are now competing with high

background noise levels in the sea (mainly the low frequency roar of ships' propellors), the call of the finback whale has a range of at least 100 miles. 'But whales similar in form to finbacks have been swimming through the oceans for 15 million years,' Payne points out. 'Their powerful calls must have been adapted for some purpose long before propellor-driven ships arrived.'

With Douglas Webb of the Woods Hole Oceanographic Institute in the USA, Payne calculated the potential range of a finback call in the quieter oceans that no longer exist and came up with a figure of 500 miles! 'This means that a finback could have a companion within 800,000 square miles of ocean,' he concluded. It could be even further away because in the deep ocean speed increases with temperature and pressure, causing sounds to travel faster near the surface and at the bottom.

'According to our calculations, if a finback were at optimum depth, prior to the intrusion of ships' noises, the maximum range of its call would be between 4000 and 13,000 miles. A circle with a 4000-mile radius covers about 50 million square miles – some 18 million more than the area of the Atlantic Ocean.'

What he is suggesting is that when the whales evolved their low-frequency calling techniques, every finback in the world could have been maintaining contact with every other. If you consider the finbacks as a network (a series of relay stations) and feed into that network the calls of all the other whales, and perhaps the echo-information being gathered and transmitted by the small whales and dolphins, there is a sense in which the seas become a grid of cetacean communication.

Relating Roger's findings with finbacks to the two forms of humpback sounds, the Paynes came to the general conclusion that the humpbacks used the low frequency calls to convey simple messages to more distant creatures: 'They could be serving as a beacon that gives the listener some sense of distance, depending on how much is heard'; and that the complex high-frequency repertoire was for listeners nearby.

What is the function of the high-frequency songs? Rather disappointingly they discovered that the omission of phrases and the other kinds of song changes bore no relation whatsoever to the time of year: the humpbacks changed their songs all the time and did so without any detectable regard for anything going on around them. They sometimes

CETACEANS – Whales and Dolphins
Cetaceans (which, despite their fishlike shape, are mammals) divide into two large sub-classes, based on the different ways of feeding.

1. The *Mysticeti*, all of whom are large whales, sift plankton (marine micro-organisms) through a whalebone (baleen) filter, and are toothless. This group includes the largest living creature ever to inhabit the Earth, the Blue Whale (*Balaenoptera musculus*). There are nine other species all inhabiting the deep ocean, ranging in size from the rare 6 ft-long pygmy right whale to the 90 ft blue whales.

2. The *Odontoceti*: All have teeth and hunt other sea creatures like fish, pteropods, cuttlefish, seals and in the case of just one species, the Killer Whale (*Orcinus orca*), other small whales. This is the most successful class of cetaceans, with 76 species amongst 38 genera, and it includes all the dolphins as well as many whales.

Cetaceans are found in fresh and salt waters. The river dolphins (the Ganges, Indus, Whitefin (China), Amazon and La Plata dolphins) are regarded as the most primitive cetaceans, because of their smaller brains and elongated snouts. Their existence has been the mainspring for the theory that cetaceans may have evolved initially in swamps and rivers before making their way into the seas. The river dolphins have extremely poor vision and some of them are only able to detect night from day.

ROGER PAYNE
Whales

made changes rapidly, sometimes slowly. They even changed some themes rapidly while they held others fixed. 'The complexity of what they were doing became more and more impressive, the deeper we looked.'

Their purpose may be pure aesthetic pleasure on the grand scale – the equivalent of what a human audience would feel in response to a massed choir which has worked for months to perfect Verdi's *Requiem*. Or it could be more simple, just everyone humming the latest popular tune. It could even be basic pleasure with a purpose, part of the mating preliminaries: emotional triggering in much the same way as we create the 'right' atmosphere with soft music.

Some of the more recent work on the songs by the Paynes' follow-up team, particularly Peter Tyack, would indicate singing is probably an aspect of court-ship, together with aggressive male behaviour. 'If so, humpback bulls are like warriors of old,' Roger Payne believes. 'They have to demonstrate skill in the arts as well as with the sword to win their ladies.'

There could, however, be a more intriguing possi-bility. Long before the ability to write became common, history and information vital to the survival and expansion of the human race were recorded in the forms of rhythmic prose and songs – the Norse sagas, for example. Perhaps every year whale 'his-tories' are updated and the information refined, as if a new and exquisite panel were being added to a masterpiece that is also a document, such as the Bayeux Tapestry. In every sense it would be a vast fabric. 'If a male humpback sings for all his adult life, think of the number of songs he must have mastered,' Roger points out.

Every animal has a need to pass on information to its fellows and its young. Every year we humans update our histories as new data becomes available, and we do it as a team. Darwin's theory of evolution is not an entirely original 'song'. Had he been a whale the only way he could possibly have refined its various elements, incorporating as it does the thoughts and expressions of many nature-watchers before him, would have been by gathering his peers around him, sounding out their opinions and hearing their evidence. Indeed that is exactly what he did, in correspondence and conversation.

At first glance this makes the 'histories' the whales are composing seem very basic – a collation of the knowledge acquired on their migration – but that is not necessarily so. We know that the whales have huge brains and excellent memories. But we have no idea what a whale sound means and it is entirely possible that the extension of a sound and the alteration of its position in a song could convey an enormous amount of information. We do it with our language – take the two words 'tack' and 'attack' – there is only a breath of difference in the sound yet a world of difference in meaning.

Humpback and other animal songs
At the very least, humpback songs may be important pointers to the evolution of singing by animals. The Paynes, with Peter Tyack, compiled this comparative chart.

continued

9. Vocal communication, which is very elaborate in its structures in some species.
10. Seemingly inquisitive natures, particularly in dolphins, which has brought them into contact with Man throughout history.

Singing Type	Representative Species	Mechanism of introducing variability	Result
1	Cricket	Genetic (mutation or hybridization)	Rigidly fixed songs with little variability
2	Chaffinch, White-crowned Sparrow	Learning from members of same species during critical period	Modest song repertoires
3	Winter Wren, Mockingbird, Brown Thrasher	Learning throughout lifetime (from other species as well as members of same species)	Larger song repertoires
4	Humpback Whale	Learning of modified versions that were derived little by little from existing songs.	Rapid continuous song evolution: modest song repertoire at any given time but almost limitless repertoire over many years
5	Human	Learning *de novo* compositions governed only by laws of form.	Rapid discontinuous song evolution, largest and most varied repertoires

ROGER PAYNE
Whales

Humpback sounds and population studies
Collecting whale sounds has become a hobby for
whale-watchers all over the world in the wake of the
Paynes' pioneering work, and these are already providing general information about whale migrations
which could be of great value to their conservation.

Roger Payne recalls receiving a set of tapes from a
group of islands near Baja, California, which he and
Katharine initially thought had been mis-labelled
because they displayed the familiar patterns and
variations peculiar to the whales they were hearing in
Hawaii. Until then it had always been assumed that
humpback herds that wintered in different locations
also travelled to different summer homes, Hawaiian
humpbacks going to the Aleutian Islands, Baja humpbacks to southern Alaska.

One of the Paynes' co-workers, Jim Darling, had
built up a portfolio of the distinctive tail markings
of the humpbacks of Hawaii. He showed them to
Charles and Virginia Jurasz who have been watching
the humpbacks of southeast Alaska since the late
1960s. The Juraszs immediately identified seven
Hawaiian whales. 'The discovery revolutionized our
entire concept of humpback populations and their
movements,' Roger admitted. 'Humpback populations
have recovered significantly since the days of whaling
when there were separate kill quotas for each breeding
area, based on the belief that there were two or more
separate breeding populations in each ocean. Now
we believe the humpbacks in each ocean mix freely
on their breeding grounds and if significant hunting
were ever to resume, this new knowledge could prove
invaluable.' He also points to the fact that if there were
a decline in one breeding area it might be the first
indication of an ocean-wide decline.

When the Voyager satellites took off in 1977 for far
galaxies with their package of songs from his beloved
whales, the light of hope appeared at last to be shining
at the end of the long tunnel, and Roger Payne felt
able to give vent to his true emotions. 'The songs of
whales, so long confined within the seas, have welled
up out of their ocean prison, overflowed the rim of
the sea, conquered the hearts of their age-old enemy,
Man, and are now on board a spacecraft bound on a
1.2 billion-year voyage that will spread their message
throughout the galaxy,' he recorded jubilantly. 'It is,
in fact, a wonderful boast to another civilization. It
tells them that by the time we sent this message we

had matured enough to give the culture of another species a bit of room on board.'

The right whale

Midway through his five-year preoccupation with the humpbacks, Payne heard of the plight of the southern right whale and he set up a conservation-orientated study programme on this endangered species in Patagonia (a region of Argentina). The study is still under way.

History of right whale hunting

The right whale is ironically named. 'It is right for all the wrong reasons,' Roger observed. 'Slow enough to catch, fat enough to float after the kill, with a treasure of whalebone in its jaws. The stakes of the hunt were incredibly high: it cost about US $6000 at the turn of the century to equip a whaling ship, yet a single right whale produced about US $12,000 in whalebone, plus the profit from its oil.'

Below: *The right whale for all the wrong reasons. A member of one of the two last remaining schools of right whales blows off the beach at Punte Norte, Peninsula Valdes, Argentina. Now totally protected in these waters thanks to Roger Payne and local activists, right whales are still, after the bowheads, the most threatened of the large cetaceans.*

ROGER PAYNE
Whales

These commercial whalers were bringing primitive technology into play in the final war of attrition on a creature that had been hunted for six centuries. There are accounts from the twelfth century of right whale hunts from open boats launched from beaches. Indeed, right whales sometimes approach so close to the shore they have been speared from promontories.

These ancient whaling stories show that the right whales once thrived off the east and west coasts of Australia, Africa and North and South America. The Pilgrim Fathers landing in Massachusetts Bay in 1620 were delighted to spot 'the right whales' swimming in the bay itself.

The journal of the *Mayflower* contains the following entry:

'We saw daily, great whales, of the best kind for oil and bone, come close to our ship, and, in fair weather, swim and play about us. There was once one, when the sun shone warm, came and lay above water, as if she had been dead, for a good while together, within half a musket shot of the ship; at which two were prepared to shoot, to see whether she would stir or no. He that gave fire first, his musket flew to pieces, both stock and barrel; yet, thanks be to God, neither he nor any man else was hurt with it though many were there about. But when the whale saw her time, she gave a snuff, and away.'

From the harpoons that followed the trajectory of those bullets, some of America's oldest fortunes were made. An indiscriminate slaughter began off New England, progressed to the huge right whale populations of South America, and finally reached out to Australia, New Zealand and Kerguelen Island where, in the middle of the nineteenth century, 14,000 right whales were killed each year.

By the close of last century the whaling industry appeared to be sharing the extinction of its prey until in 1901, a chemical technique for removing the unpleasant taste from whale oil was developed, making it possible for it to be used in food products.

A new wave of twentieth century technology – the steam ship and the harpoon gun – was employed as the whaling industry experienced a new, and even rosier, dawn. The easy whales, 'the right whales', were quickly reduced to a rarity but the faster, more seaworthy ships, armed with inescapably-deadly explosive harpoons, proceeded to decimate the finbacks, the seis, the blues and the humpbacks. This went on for 30 bloody years until the principal whaling nations, in the interests of husbandry rather than conservation, agreed a limp treaty in 1930 for the protection of right and gray whales (there were no international observers to monitor its effectiveness then). The gray whales made a slow but steady recovery.

Not so the right whales. It was as if the slaughter had robbed them of the will to survive, and 40 years after the 1930 treaty, with no serious renewal of the attrition, right whales were, as Roger Payne puts it, 'hanging on by the thinnest of threads. To see one at all is a privilege and a major event in anyone's life.'

In 1970 Roger Payne, supported by the New York Zoological Society (whose reputation for rescuing endangered species began with the American bison), set out to find out why these mammals were so scarce. Fifteen years later he is still watching, but at least some aspects of the enigma of the right whales' death-wish have been solved, and there is still 'a small colony of these playful and unaggressive creatures in an isolated corner north of the Straits of Magellan (Chile), far from the predatory greed of Man.' (There is one other, possibly larger, colony off South Africa.)

'To find myself on the beach of Punta Norte on Peninsula Valdés (Argentina) was the realization of a life-long dream' Roger recalls. 'It was dotted with elephant seals, spaced as regularly as beads on a string, while patrolling just off shore at a slightly wider spacing were six right whales.'

Thus began the first, and to our knowledge the only behavioural study of right whales that anyone had conducted, even though our busy persecution of these creatures has been continuing for centuries.

Behavioural studies on right whales

The right whale, after the bowhead whale, is still the most endangered whale species, and until Payne began his studies no way had been found to assess their condition and their biology properly without doing further damage, which was an unthinkable proposition given their desperate rarity.

He began by developing methods for counting and identifying the whales, avoiding anything resembling the former practice of firing marked bolts into the whale, which were later recovered from the blubber kettles of the factory ships! By noting differences between natural features of individuals, Payne has been able to record the return of identified right whales year after year. In short, he has made the first

Above: *Roger Payne now eavesdrops on whales worldwide: here the calls are heard off Negonbo, Sri Lanka.*

ROGER PAYNE
Whales

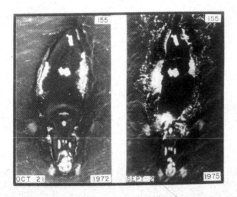

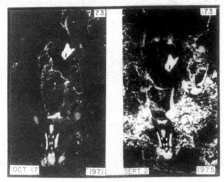

Above: *Aerial photography has revealed to whale-watchers that the leviathans have distinct, individual markings allowing much more accurate observation of behaviour.*

extensive study of the natural behaviour of a society of free-ranging baleen (plankton-feeding) whales, and has gained insights into the social forces holding a population of whales together.

The work is done largely from aerial photographs: 'By repeatedly photographing a herd, you learn to recognize familiar features of individuals,' Roger explained. 'You can sort out the strangers and, by determining the ratio of familiar to unfamiliar faces, estimate the size of the population. One can likewise determine friendship and kinship groups and thus learn how social behaviour contributes to reproductive success.'

In addition there are the 'magic moments' when new, sometimes bizarre, behaviours turn up. Roger recalled the first time he saw one of these huge creatures standing on its head. 'We see a lot of that,' Roger confirmed. 'They are actually "sailing" on the currents. More rarely you will see one with its tail out of the water, nose down, eyes closed, with its mouth inches off the bottom.' It has been suggested that this is a kind of feeding behaviour, but Roger is inclined to the view that the whales are simply taking a nap.

'I've seen them stay up-ended like that for as long as 20 minutes with their eyes closed, when normally they would blow (rise to the surface to breathe) every five or six minutes.'

The whales are measured by photographing them alongside a boat with a one-metre disc hanging over the side. Overall length is easily gauged using the disc as a scale.

Payne's techniques provide ways of measuring the basic parameters of a population of whales, including age, sex, rate of growth, reproductive condition, calving rate, recruitment rate (number of individuals added to the herd), mortality, and population size, as well as whether a population is increasing or decreasing. All this can be determined without marking or molesting the whales in any way. Others have learnt from his methods, and similar techniques are now used for the study of humpback, gray, finback and sperm whales.

The most gratifying result of all these years of careful study is to give a possible reason for the right whale's painfully slow recovery rate. Payne and his students have established that most female right whales only give birth once in every three or sometimes four years: a calving rate that seems to be slower than any other species of large whale.

All that means, however, is that the future for right whales is still in doubt. As Payne once wrote: 'In the late spring, Patagonian right whales leave their breeding area – an area protected by the Argentines – and head south to summer feeding grounds. But where do they go? Will they evade the guns of whalers who are party to no restrictions?'

Conservation measures

Haunted by such nightmares, Payne has devoted as much time to the whale conservation movement as to his own studies of whales. He is an activist within the IWC and a travelling preacher in the whale cause, lecturing to any audience that will hear him. Although the most recent meetings of the Commission have resulted in ever smaller 'quotas' of permitted killings and a moratorium on hunting (which alas is already showing signs of breaking down), rogue whalers still ply their bloody trade and countries with large whaling fleets are switching their resources to dredging krill, the tiny plankton on which all baleen whales depend for food. This is arguably a more dangerous development than the very limited supervised whaling which is all that has been sanctioned in recent years.

Payne's sponsors during his Patagonian research, the New York Zoological Society, created a special

Below: *Suspended in a cocoon of blue space an adult sperm whale keeps close company with a newly-born calf. More than three-quarters of the Earth's surface is sea and nature-watchers consider that this vast space has probably shaped the whales mentally as well as physically. They are, in a sense, more in tune with space creatures than land-dwellers.*

ROGER PAYNE
Whales

fund, enabling him to campaign that all the species of great whales should be added to the United States Endangered Species list; this bans the importation of either the parts or products from such species. The ban was enforced and it has since been estimated that this single piece of legislation reduced the potential market for whale products by US $35 million a year. The US Marine Mammals Protection Act followed soon afterwards (see 'Whaling', p. 48).

In 1974 Golfo San Jose on the coast of Argentina in the province of Chubut, next to Peninsula Valdés, was set aside as a reserve for whales in perpetuity, with all commercial exploitation and development prohibited, following a series of meetings with officers of the Province of Chubut.

California's somewhat bizarre Whale Day in 1976 was the product of a meeting between Governor Jerry Brown and Roger Payne, but it did have a powerful spin-off. The state legislature awarded a grant of nearly US $1 million for the experimental growing of jojoba nuts, which yield an oil equal to or better than sperm whale oil. The oil has been a huge commercial success, thanks largely to its use in shampoo, and no American manufacturer would now contemplate using sperm whale oil in preference to jojoba!

Payne has also served on a number of United Nations committees seeking whale protection: he has drawn up the guidelines for the Maui County Whale Reserve in Hawaii, which was another slightly bizarre, though essential conservation project, in that its purpose was to protect the whales from harassment by tourists.

In 1978, Roger Payne was made a Knight of the Order of the Golden Ark by Prince Bernhard of the Netherlands. This is the premier accolade of the international conservation movement, and was awarded in recognition of his 20 years work on conservation campaigns for whales. Recently he became the research scientist to the United States World Wildlife Fund, and this is now his main job.

Payne believes that the future of whales will only have been secured when public consciousness has been raised to a point where their slaughter is reviled as the murder of a sister species in an adjoining habitat, an equal neighbour and vital partner in a civilized existence. He knows, as well, that the world is still ethically and economically far removed from such a state of grace, and that such a state may only be constructed out of respect, admiration and a sense of wonder.

LAND

4. TREE-WATCH

Roughly 400 million years ago a most dramatic migration occurred: life moved from the oceans onto the land. Plants colonized the planetary surface long before any animal trod those ancient soils and so it seemed sensible to begin our land-watch in the plant kingdom.

Blue algae and other unicellular water plants had created the basis for the existence of animal life in the seas and lakes of the primitive earth, and the same was to happen on the land.

We know this to be the case because of the fossil records of plants, which, while not as spectacular as animal records, are still comprehensive. In the Swiss quarry of Ohningen, for example, where fossil animal finds revolutionized eighteenth century attitudes towards evolution, 475 fossil plant species were also found. Since then we have found fossil algaes, mosses, grasses, weeds, flowers, pollen, trees, resin and even whole fossil forests.

Admittedly the record is far from complete and we cannot pretend that all the steps in plant evolution are known. On the best evidence available it seems to have happened as follows.

In the primal soup of the Earth's first waters some 3.5 billion years ago, archaic bacteria, the first organized living things, developed; to be followed a million or so years later by the pre-Cambrian 'uralgae', the first oxygen-producing plants. Very little is known about either of these two groups of early plants.

We suspect, however, that because the first single-celled blue-green algae were successful plants in their own simple way, their descendants which still flourish today are probably very similar.

Fossil algae do exist, but because of the delicate structure of these organisms such fossils are rare and give no true indication of the processes of change and selection which eventually led, eons later, to the appearance of more 'plant-like' plants in the damp transitional zones of marshes and sea shores. These first migrants to the land had stems, and the fossil record reveals their existence 440 million years ago in the Silurian period, with evidence of extensive colonization over the next 40 million years.

The Fossil Hunt

Fossils – animal or plant remains or their outlines preserved in rocks – are probably the first form of natural history collecting of which we have definite evidence.

Excavations of Stone Age sites at Peterfels near Vienna indicate that primeval hunters wore these mysterious stones as ornaments, and more than 50 Tertiary fossil shells were found in a Bronze Age urn buried at Benherg in Thuringia. Ancient Chinese legends describe invertebrate fossils as the remains of dragons, and in 1200 AD the poet Shushi wrote: 'Once I saw shells in the rocks high in the mountains. I am quite certain that they were the shells of marine mussels. Thus the rocks must once have been ocean.'

In fact the scholars of ancient times, uninhibited by modern religion (which insisted that everything on Earth was the week-long work of God) were more aware of the true nature of fossils than later scientists, who puzzled over their origins, right up to the nineteenth century. Greek scholars of the sixth century BC like Empodocles and Herodotus believed fossils were the product of petrifaction.

For centuries thereafter, however, the stones were generally regarded as the whim of the gods, and their bizarre distribution as evidence of playful creation, or of Noah's famous flood. Around the year 1500 AD, however, the remarkable Leonardo da Vinci considered the 'Deluge' theory of fossil distribution – and quickly rejected it. 'If you think the Deluge carried these shells many hundreds of miles from the seas, I reply to you: that cannot be. For the Deluge was caused by torrential rainfall which the rivers naturally carried to the sea, along with the dead things that had been washed into them.

continued

Opposite: *The 'highly-strung' Ron Stecker keeping a nature-watch on the most massive living things on earth, the giant sequoias in California.*

continued

The rainfall did not . . . draw the dead things from the shores of the sea to the mountains . . .' Some 30 years later he was being supported by the German physician, George Bauer (Agricola) who named the stones 'petrifactions', and later 'fossils' (from the Latin *fossilis* dug up). However, a brave rebel, Bernard Palissy, who dared suggest in the 1560s that 'shell-stones' were the remains of sea creatures and that central France was once a bay in the ocean, was charged with heresy and died in the Bastille.

It took the protection of the Medicis a century later to provide a platform from which a scientist, Niels Stenson, known as Steno, dared re-state what many now saw to be obvious. Steno studied *Glossopetrae* – 'tongue stones' – and decided they were not celestial baubles or natural curiosities but the teeth of long-dead sharks.

The dam of dogma broke finally when the wealth of fossil finds defied explanation. There was a small 'gold rush' to the quarries of Ohningen in Switzerland in 1775, when masses of rare fossils, including elephant tusks and crocodiles, were unearthed: they found a ready market in the collections of the rich, including many a crowned head. There was even a thriving trade in hand-carved forgeries.

Over the next 50 years the true 'catastrophes' largely responsible for the distribution of fossils, the ice ages, were revealed by scientists like the Comte de Buffon, Jean Baptiste Lamarck, Alexander von Humbolt, Goethe and Louis Agassiz.

Known as the sporophytes, they were naked, leafless and oddly-branched. They were, however, infinitely more advanced than the algae, with 'internal (vascular) plumbing', and sophisticated reproduction by means of microscopic spores in sacs on their stems or leaves. They paved the way for plants of similar design; ferns, horsetails and club mosses prevailed on the earth for some 50 million years thereafter.

Then came the period we know as the Carboniferous, when the whole world was a kingdom of plants; swamps thick with undergrowth, luxuriant jungles, verdant montane forests. The remains of this explosion of vegetable life formed the huge carbon (hence 'Carboniferous') deposits we call coal, or rich beds of humus to fertilize countless succeeding generations of plants. One of the great mysteries of the evolution of the Earth is why the Carboniferous was so wet and the preceding period, the Devonian, so dry. It seems likely, as with modern rain forests, that once the early plants had established a foothold on the land, a rain cycle was sustained. An ever-increasing greenhouse effect promoted more plant growth until a vast tropical jungle was created over the entire land surface of the earth.

What did these plants look like? If we enlarge contemporary ferns, horsetails, and club mosses to the size of trees (there are still 65 species of palm fern living in South America, Africa and Australia), we would have a reasonable facsimile of the forests of the Carboniferous.

During this period some plants also progressed along a path of selection which was to produce a dramatically more efficient method of reproduction; the blueprint of the system still employed by the most advanced of today's plants. Spores are very inefficient; of the millions produced only a handful fall on well-lit, moist ground which will promote germination and the formation of sex cells which, after combining, will give rise to a new club moss or tree fern.

The improved system was the seed, a life-support capsule of food round an embryo or partially developed young plant, cased for survival in a protective shell.

Living fossils

Unlike the dinosaurs of the animal kingdom who have vanished completely into historical record, several of the most primitive seed-producing plants still exist, seemingly almost unchanged, today.

One such is the gingko (or maidenhair tree), seedlings of which reached England in the early 1700s from China and Japan and were immediately spotted as a rarity by Western naturalists. Charles Darwin, who saw one at Kew, termed it a 'living fossil'.

Gingkos have the oldest fossil record of any plant on earth and are related to the earliest surviving seed plants, the conifers, who have adapted so well to various conditions (from arctic tundra to scorching deserts) that they have varied little in the 350 million years they are known to have existed.

They owe their success to the evolution of cones, a hard stack of spore-bearing scales. On these scales (hard leaves) one or two large spores (macrospores) develop, which, after fertilization, give rise to seed. Fertilization is achieved by the pollen produced in male cones being wind-borne to the larger, seed-bearing, female cones. From this wind-dependent method of fertilization, other plants went on to a more successful method yet: insect pollination. To attract insects, plants had to evolve a lure, their flowers. They also had to wait for the evolution of insects and these are matters we deal with elsewhere.

Below: *The last refuge of the living dinosaurs: giant sequoias, trees with a continuous record dating back to the era of the dinosaurs are now confined to a number of groves in Western America like this Redwood Mountain Grove in California. Nowadays 95% of American giant sequoias are protected and thriving.*

RON STECKER & TOM HARVEY
Giant sequoias

Above: *Tom Harvey: getting to the core of sequoia life cycles.*

Below: *King's Canyon National Park, California.*

For the moment we will remain with the ancient giants of the Earth, the conifers, and in particular with a group of trees that we may rightly regard as living dinosaurs – the giant sequoias. They are the most massive living organisms on earth with a continuous record dating back more than 125 million years. Huge forests of sequoias once existed in many parts of the world, including Europe; today they are confined to a number of groves in western America. These groves are believed to be the last ecological refuge of a much larger forest of sequoias established in South Idaho and Nevada. Cool, dry conditions descended on these areas when the great upheaval which created the Sierra Nevada range occurred, and the sequoias migrated westward through the lower mountain passes to their present locations in California.

The trees that grew up in the new western habitat found conditions to their liking at an altitude between 5000 and 6000 feet on the western Sierra Nevada where the air is cool and moist. Seventy-five groves survive today covering 35,000 acres. This may seem like a lot of trees, but in terms of the long history of the species, it could be their final refuge.

The survival of the giant sequoias is, instead, one of the great successes of world conservation. Today 95 per cent of all giant sequoias are under some type of State or Federal protection, a total of well over 150,000 trees. We visited the largest of several conglomerates, the Redwood Mountain Grove in King's Canyon National Park, California, where 70,000 trees live out their lives in absolute sanctuary, to discuss their past and their future with two tree-lovers: Ron Stecker and Tom Harvey.

RON STECKER AND TOM HARVEY
Among the giants

The health and welfare of the trees is monitored with reverence by a visiting brigade of scientific naturalists. As with any visitor to the Grove, Ron Stecker first set out to imbue us with a suitable sense of scale.

Vital statistics of giant sequoias
The facts, of course, are impressive enough on their own. Giant sequoias reach a height of some 300 feet, have trunks 30 feet across at their base and can live for 3000 years. It is no wonder that they attracted ferocious commercial interest when you consider that one tree would produce 3000 posts, sufficient to fence

Giant sequoias

an 8000-acre range or 650,000 shingles, more than enough to roof 70–80 log cabins. At the other end of the scale are their seeds. Giant sequoia seeds are the size of oatmeal flakes; it would take 100,000 of them to make up a pound in weight!

Ron Stecker's research has produced a mass of intriguing new information, much of it to do with dynamics of sequoia reproduction. In a normal year the tree which Stecker studies, called the Castro Tree, produces 1400–2000 new cones (most large trees have an average cone load of 11,000). Cone production begins when the tree is 10–14 years old, and can continue for 3000 years. The average germinability in seeds in mature cones is about 35 per cent. From the 2000 cones an average of 200 seeds are released, giving a seed yield of 400,000 per tree.

At first glance the success rate of the giant sequoia's seeds looks miserable, but Ron Stecker reminds us that it is necessary to view this reproductive rate in the context of sequoia longevity: 'The Castro Tree is 1500 years old, others around it are 2500 years old. How much reproduction do you really need? If two succeed per acre every 100 years you would probably have too many.'

The trees live a life of three distinct phases. Sapling trees develop conical 'spire tops', and where sequoias are virtually the only species, they make dense stands. Spire tops may persist for 100 years or more when the light is good and these young trees can achieve trunks with a diameter of four feet (at breast height), with several inches of bark. When they reach a height of 300 feet, vertical growth slows in comparison to the growth rate of the lateral limbs and at around 600 years the crown gradually assumes the broadly-rounded form which is the trademark of the giant sequoia. The trunk of a well-formed mature tree is 10–20 feet in diameter. In old age, the trees develop a 'snag top'; a dead section of tree usually caused by fire damage interrupting the flow of water and minerals to the high crown. The older they grow, the more they lose their symmetry and grace of youth.

Surprisingly for so large a tree, sequoias have shallow root systems reaching down no more than 8–10 feet below the ground. The spread of roots is dictated by what is called the 'drip zone'; a fully grown tree will have a fine net of roots radiating out 100 feet or so, fed by drips from the canopy.

Very large sequoias can grow as close together as ten feet which has caused the experts to believe that

Trees

A 'tree' is a mode of growth, not a botanical class of plants.

Trees number among the two classes of flowering plants, the gymnosperms and the angiosperms. Conifers and ancient trees, like the gingkos, are gymnosperms (plants having 'naked seeds' on a bract), while more 'modern' trees, like the elm and lime, are classed with the angiosperms because they have seeds hidden in an ovary.

The feature which properly defines a tree from other forms of growth (except shrubs) is that it makes 'wood': a single layer of cells called cambium. The outer cells of the cambium (called phloem) conduct food manufactured in the leaves down to the roots, while the inner cambium (xylem) conducts water and liquid nutrients upwards from the roots, where they have been extracted from the soil.

Trees are quite properly classed with the flowering plants because they all have flowers, but only those that rely on pollinators have obvious ones (with petals). In the course of their evolution, trees have explored a number of systems of propagation. The most ancient are trees where male and female organs are on separate trees (dioecious – two households). Some modern trees (such as the willow) have reverted to this arrangement but the majority are monoecious, employing the more efficient technique of male and female flowers on the same tree, or in the same flower.

RON STECKER & TOM HARVEY
Giant sequoias

they hold each other up. They also have a huge mass at the base of their trunks and have been compared to toy Russian dolls which have counterbalances in their bases to provide flexible stability.

Giant sequoias and fire

Sequoias have an almost symbiotic relationship with fire. When the conservation of American forests was begun in earnest some 50 years ago the prevention of forest fire was regarded as the first priority; for most forests this measure saved many tree lives. But nature-watchers concerned with sequoia forests began to question this policy when they became aware that apparently healthy sequoias were not casting their seed.

As Tom Harvey recalled: 'Both the loggers, and later the conservationists, had been keeping fire out of these forests for almost a century when we first suggested that maybe this was the wrong policy for

continued

Level Two is inhabited by creatures that get their energy from plants – 'herb' eaters (herbivores) like insects and squirrels.

Level Three is the domain of creatures who get energy from both plants and also the odd plant-eater. Hedgehogs and a lot of birds are both herbivores and carnivores.

And finally in *Level Four* are the true carnivores, like the owls and foxes, who get their energy from an exclusive diet of meat.

A simpler view of this 'trophic pyramid' as it is called, is to see it as a food chain – a plant taking up soil nutrients is eaten by a caterpillar which is eaten by a pygmy shrew which in turn is eaten by a kestrel.

A food 'web' is an even better way of describing what goes on. We suggest you get started on your foster tree and its surroundings by filling in who's living where, and eating whom, at each level of the energy pyramid. To do this, and to create a permanent record of your foster tree, nature-watchers have evolved a number of techniques; one of them is tray beating.

Tray Beating
Make a tray out of a piece of white cloth tacked over a board, and align it beneath a branch. With a stick, give the branch a sharp rap. Lots of little twig-eaters and leaf-eaters will get the shock of their lives and tumble onto your tray.

The kind of nature-watch we propose does not require that you imprison the subjects, so when you have made notes

continued

Above: *A giant amongst giants – the General Sherman sequoia, the largest living thing on earth.*

Opposite: *Fire is the key to giant sequoia survival; burns (now carefully controlled) clear and mulch the forest floor, while the heat induces cones to open and shed their seed.*

sequoias. Eventually we were able to convince people that giant sequoias need fire.'

Fire, it seemed, had been doing the housekeeping in the forest for all those countless years before the present breed of guardians came along.

In 1969, Tom and a group of tree-watchers witnessed a fiery 'surface burn' through the Redwood Grove. The burn eliminated most of the plant life on the forest floor except for the giant sequoias. They were protected by their insulating outer bark. The other trees, like white firs, which compete with the sequoias for light and moisture in the forest, were culled out.

'A few days after such a fire, the closed cones in the upper branches of the giant sequoias open and the

RON STECKER & TOM HARVEY
Giant sequoias

continued
and taken photographs, put your insects back into the tree.

Climbing about in trees can be dangerous, and a safer technique is to choose a low hanging branch, spreading a sheet on the ground beneath. Again, shake the contents of the sheet back into the tree when you have finished.

Be warned that fostering a tree is no mean task. Most people find a whole tree ecosystem too detailed for a full study and choose to focus their attention on one particular aspect.

Bark Rubbing
You can keep a good record of your favourite tree, and create an unusual portfolio that will help identify other similar trees, by 'rubbing' a section of trunk just as some people 'rub' old brasses in churches.

Tape a square of waxed or grease-proof paper to the trunk and rub away with the flat side of a wax crayon. Always work in the same direction.

Pressing
Pressed dried leaves can also go into this portfolio for identification, or just simply for their delicate translucent beauty.

The first rule is to take care transporting the leaves home; the simplest method is to press them between the pages of your field guide. There is a special case, called a vasculum, designed for this purpose by nature-watchers, but a plastic box of the *Tupperware*-type does just as well.

The trick is simply to keep the specimens under heavy pressure until they are dry. Custom-made presses are not expensive. You can make one out of an old tennis racquet press and two squares of hardboard. Excellent results can be obtained from an old tie or clothes press; these can still be found cheaply in antique shops.

To get the best results, clean the specimens in water (taking care to re-
continued

forest floor becomes littered with seeds,' Tom explained. 'In the cleared, mineral-rich mulch it will only take a little rain for a tiny giant sequoia to spring up.' So the sequoia population benefits from forest fires which weed out competing trees and provide the right conditions for seed germination.

Tom admits to being a forest-watcher of the old school, conditioned to a deep-rooted horror of fire that he has found difficult to overcome. But he points out: 'Within the destruction there is a rebirth – a phoenix that will spring forth from the fires. It's probably also true that fire has been affecting the life of plants ever since they started to grow on land millions upon millions of years ago.' (Lightning was responsible for these natural fires.)

The General Sherman Tree
The giant sequoia 'General Sherman', has the unique status of being the largest living thing that exists on Earth today. Its weight has been accurately estimated at 1385 tons. The General Sherman is at least 2000 years old, whereas the oldest giant sequoia is known to be over 3000 years old.

The Largest Living Thing on Earth – *The General Sherman Tree*	
Estimated age	2500–3000 years
Estimated weight of trunk	1385 tons
Height above base	274.9 feet
Circumference at ground	102.6 feet
Maximum diameter at base	36.5 feet
Diameter 18.3 m above ground	17.5 feet
Diamter 54.9 m above ground	14.0 feet
Diameter of largest branch	6.8 feet
Height of first large branch	130.0 feet
Volume of trunk	52,500 cubic feet

Most sequoia experts believe that the General Sherman owes its size to some ideal fire that cleansed the forest at the dawn of human history. The older, smaller trees wasted their youth struggling through the competing foliage of a forest that had not enjoyed the benefit of a suitable fire.

Canopy dwellers
High in the canopy of the giant sequoia forests there are ecosystems within ecosystems. The giant sequoia called the 'Castro Tree' has been accommodating a visitor, an insect specialist called Ron Stecker, for a

number of years. Stecker is ably supported by his wife, Philippa, who also loves the trees despite the fact that she fears constantly for her husband's safety.

To come to these intimate terms with this particular sequoia involved finding two trees close together and using one for access. It sounds simple: 'You spur up the main trunk, come back down on a swing line, then swing across onto the first main branch,' said Ron.

In reality, given the great heights involved, you need all the skills of the mountain climber to indulge in this form of nature-watching. The first main branch on a mature sequoia rarely occurs below 100 feet. In the case of the Castro Tree it is 140 feet off the ground, and that is less than halfway to the peak of the canopy where Ron likes to spend most of his time.

He used to worry about the work getting dull, but after a while he built up a very special relationship with the Castro. To find his way around, he named each of the branches of the tree after people, and in time found himself talking to them as if they really were people.

RON STECKER & TOM HARVEY
Giant sequoias

continued

move any insects) and lay them between a couple of sheets of blotting paper. You can press several specimens by putting them in layers if you have a good-sized press; you should then use three sheets of paper between each layer.

Printing and Stripping
Delicate prints can be made from leaves by coating them with a colouring agent (paint, coloured varnish, boot polish, or a dusting of graphite).

Place the coloured side down on a sheet of good quality paper and press gently using a sheet of blotting paper. It is also possible to decorate furniture in this way. These are delicate techniques, and you can achieve stunning results by varying the colours on the leaf and the print surface. Practice first.

Simpler representations of leaves that are also useful for identification can be made by the silhouette technique. Simply lay the leaf on a sheet of paper or board and spray it with an aerosol paint. A careful choice of colours and backings can produce a very artisitic silhouette. Regard your portfolios as an aesthetic record.

Filigrees
Sometimes on walks in the woods you will see leaves that have been reduced to a delicate filigree by decay or insect attack. A filigreed leaf can be created for your portfolio by simmering the specimen in water for an hour, leaving it in the saucepan with the water for a few days until the leaf flesh is soft, removing and rinsing the specimen, and etching the leaf flesh off the skeleton, using a paintbrush dipped in pure bleach.

Left: Ron Stecker taking insect samples high up in the canopy of a giant sequoia.

RON STECKER & TOM HARVEY
Giant sequoias

Ron Stecker studies the creatures that share the Castro's life. When he first began his incredibly detailed survey, 20 types of insect were known to live in the sequoias at one time or another. By the end of his first summer of research, Stecker had extended the count to 46.

Important to dispersal of the seed is the activity of another small creature, a beetle, *Phymatodes nitidus*, discovered by Ron Stecker in 1968.

'The female beetle lays her eggs in the junctions of the cones' scales. The larvae, measuring only an eighth to a fifth of an inch in length, chew their way into the cone's interior, obtaining nourishment from its tissues. Their tunnels, or *mines*, are packed with chewed and digested waste, which resembles a fine salt and pepper mixture. Often the vascular system

Below: *A small beetle,* Phymatodes nitidus, *is helping giant sequoias survive. Cones, invaded by beetles for egg-laying, dry and shed seed more quickly.*

of the cone is severed during this feeding, which cuts off water conduction to the ends of the cone scales. As the mining turns the green, fleshy tissues into a sun and air-dried cone, the scales shrink, creating gaps between them; the cone's hold upon the seed is relaxed and dispersal follows.

'Normally the beetles do not eat the seeds, although they may damage some of them as they move within the cone. Browned cones don't necessarily drop their seeds immediately upon opening but rather drop them slowly over the years as the vascular connections are severed and the winds shake them loose.

'The beetle larvae average 1.4 individuals per cone in those attacked, while there may be as many as eight in a single cone. Up to a third of the cones in the average tree are brown cones in which the beetle larvae have been active.

'We see that the wellbeing of these huge giants is dependent to a large degree on this tiny insect, and vice versa: it seems that each species is dependent on some reciprocal arrangement of services. In this case the tree's cones supply the food for the beetle and the beetle causes the seeds to be released.'

One of the problems facing giant sequoias is, as we have seen, seed dispersal. Ron Stecker has discovered that the huge trees are assisted in this task by small helpmates. A moth helps out with the first-year cones. Second, third and possibly fourth year cones are cut and the seed disseminated by hungry chickaree squirrels.

'Chickaree squirrels have a head for heights and a voracious appetite for sequoia cones. A lone squirrel was observed cutting 538 green sequoia cones in just 31 minutes. They can cut these cones at a rate of something like 40–60 per minute. The cones are the size of lemons; when there's a squirrel at work they come down at 90 miles per hour and you really need to watch out,' Ron told us.

Ron Stecker is well aware that he and his fellow nature-watchers are only one of a number of agents concerned with the future of the sequoias. 'I'd fight for them if necessary,' he said, only half jokingly, as our interview came to an end, and he began to climb up to the Castro Tree once more.

The biggest tree in Britain is also a sequoia, but by comparison with American trees, it is a dwarf. Less than a year after the giant sequoias had been found at North Calaveras (in 1853), a packet of seeds had

RON STECKER & TOM HARVEY
Giant sequoias

ALAN MITCHELL
British trees

Above: *Alan Mitchell: a passion for collecting trees. (For map see p. 76.)*

arrived in Scotland and were soon planted. British horticulturists were very keen to see if they could propagate the largest living thing on earth, and in those early days one-year-old giant sequoia seedlings changed hands at the incredible price of £10 each.

Britain's place in the record books of the tree-loving nature-watcher stems from our long and almost unique record as a nation of gardeners who have continually collected trees. That grand tradition is being carried on by Alan Mitchell, who has recently retired from the Forestry Commission.

ALAN MITCHELL:
An Englishman's home is his garden

Counting only real trees, not large bushes, there are only 34 native species to Britain, largely because we have only had 11,000 years since the last ice age to re-establish our domestic collection. We compensate for this with our voracious appetite for collecting, which probably began more prosaically with a need for fuel to keep out the Celtic cold. There is evidence that Bronze Age Britons brought in elms.

Today we proudly husband a national collection of 1800–2000 species of trees, with another 800–1000 varieties (species produced by artificial breeding). 'They came over in waves,' Alan Mitchell explained. 'As we expanded our empire and new places were discovered so new trees were imported. There was a great bonanza when steamships came along because they could get seedlings home quicker, and another bonanza when cold storage was perfected.'

Britain's often maligned climate is also an almost perfect one for a great number of tree species. Mitchell's 'home forest' is the Forestry Commission's pinetum in Kent. There he showed us what he believes to be the most comprehensive collection of conifers in the world, the majority of which seem to grow better here than in their own countries.

If Mitchell is to be believed, the old saying about an Englishman's home being his garden could well be true. We had heard that Alan Mitchell was more than just a tree expert, that his enthusiasm verged on eccentricity. Alan quite cheerfully admitted this: 'It's a total mania,' he said of his passion for collecting trees and information about them. 'It will keep me going for the rest of my life, I hope.' We told him he had a soul-mate up the Castro Tree!

But in reality, no one we have ever met has an ob-

ALAN MITCHELL
British trees

Species of tree native to the British Isles

Common juniper	Wych elm
Common yew	Hawthorn
Scots pine	Midland thorn
Crack willow	Rowan
White willow	Wild service tree
Bay willow	Whitebeam
Sallow	Wild pear
Aspen	Crab-apple
Black poplar	Wild cherry
Silver birch	Bird cherry
Downy birch	Box
Common alder	Holly
Hornbeam	Field maple
Hazel	Small-leaf lime
Sessile oak	Broad-leaf lime
Common oak	Strawberry tree
Beech	Common ash

The following are forms, hybrids and apomictic species of Whitebeam endemic to the British Isles:

Arran whitebeam
Sorbus arranensis
Two glens on Arran
Sorbus eminens
Wye Valley, Avon Gorge
Irish whitebeam
Sorbus hibernica
Central Ireland
Sorbus vexans
Exmoor, Culbone-Lynmouth
Bristol whitebeam
Sorbus bristoliensis
Avon Gorge
Sorbus subcuneata
Exmoor, Minehead-Watersmeet
Devon whitebeam
Sorbus devonientis
Devon, Cornwall, Ireland
Sorbus × vagensis
Wye Valley

(The Devon whitebeam is known locally as French hales.)

Left: *Heaven on earth for British tree enthusiasts, the Forestry Commission's pinetum, Kent.*

session that even approaches Alan Mitchell's mania for tree detail. Thus far he has collected 69,000 'records' (detailed notes on trees). In 1981 alone he had taken the vital statistics of 5800 trees in Britain. Three thousand of these were new trees. You do not need a computer to work out that he is visiting 100 trees a week, working from dawn to dusk.

The reward for all this frantic activity is an encyclopaedic knowledge of the trees of Britain:

'Where's the tallest tree in Kent, Alan?'

'The Disraeli planted at Erridge. Grand fir, planted 1866. Last time I saw it, it was 150 feet tall. It'll be over 160 feet now.'

'What about the oldest tree in, say, Wiltshire?'

'That would be the Tidbury Yew. It's at least 1200 years old, some people say 1800 years.'

'These trees are personal pals of mine now,' he said. 'Even if I were to just see a picture of them, from a long distance away, I would know them by their individual shapes. Most of them are individual

Above: *The horse chestnut at Hurst-bourne Priors Church, Hampshire.*

Right: *Map showing Alan Mitchell's 'Top Ten Trees' (p. 77).*

Eight major, world-class gardens for trees in Sussex:

  Leonardslee
  Sheffield Park
  Nymans
  Borde Hill
  West Dean and Roche's Arboretum
  Wakehurst Place
  Beauport Park
  Alexandra Park

Other Sussex tree-collections of great worth, some of them as notable as those above:

  Petworth House
  Cowdray Park
  Warnham Court
  Goodwood Park
  Highdown
  Ashburnham Park
  Buxted Park
  Lydhurst
  Tilgate Park

characters to me.'

Don't make the mistake of thinking that he is simply referring to their familiar statistics; these friendships go much deeper than that. 'I really think trees have a certain presence. The ordinary countryside oak trees don't move me at all, they're like lampposts. But some of the fine old oaks in this country – individual trees well known to a lot of people – really do have character.'

Perhaps we should rephrase our earlier suggestion that you should 'foster' a tree. Alan Mitchell's approach is: go out and make friends with one. To get you started, we asked him to provide us with a list of some of his best friends. We were forced to confine him to his Top Ten: the descriptions of these favourite trees are, of course, his as well, and provide an indication of how affectionate his feelings are towards them. (See Alan Mitchell's British Top Ten, p. 77.)

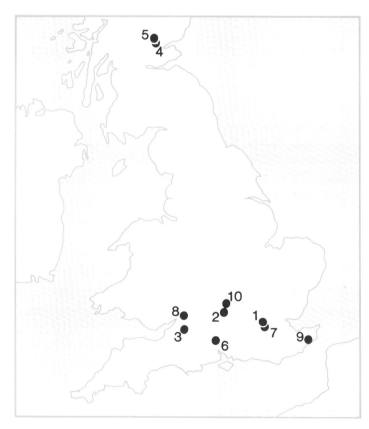

Alan Mitchell's British Top Ten

1. *Chestnutleaf Oak, Bell Lawn, Kew Gardens*
Original 1845 tree. No other example known. Always been the biggest specimen, extending its lead. Species has splendid foliage, 20cm long. Magnificent specimen 105ft × 21ft, biggest and finest oak around London and amongst biggest tree of any sort.

2. *Yellow Buckeye, Denman College, Marcham, Oxfordshire*
One of the two finest specimens of this most elegantly and brightly-foliaged tree, which has been grossly ignored and on whose behalf I (Alan Mitchell) crusade. Handily placed to demonstrate to the W.I. (Women's Institute) courses (Denman College is their national centre) the beauties of the species, and to show the plain signs of where it was grafted onto horse chestnut, leading painlessly to discussion on grafting.

3. *Chinese Necklace Poplar, Bath Botanic Gardens*
This outstanding, shapely, big and vigorous example of the species with its huge leaves, but which is usually a poor scrawny graft.

4. *Sycamore 'The Birnam Beauty', Birnam, Perth*
One of the two 'Birnam Oaks'. The biggest bole (tree trunk) of sycamore, on an old hedgebank now eroded, leaving an extraordinary wall of wood, 105ft × 24ft 5in. Leaves like rhubarb, seed like golfballs.

5. *Douglas Fir, The Hermitage, Dunkeld, Perth*
The perfect tree. Self-sown on rocks by a river in a ravine, it was about 63 years old when I first saw it, and 162ft tall. In 1957 I predicted it would be the first British tree to reach 200ft by 1990. I had not seen so many then. Now it is nearly there – I make it 197ft, but I think two or three other trees will just beat it. Beautiful, evenly tapered straight stem and light, level branches making a perfect conical crown. The foresters and treebreeder's dream. Growing as fast as any in America.

6. *Horse Chestnut, Hurstbourne Priors Church, Hampshire*
Magnificent and shapely with a clean, cylindrical bole into the crown. It is not quite the tallest, but is as wide as any, 124ft × 21ft, and typically superb in flower.

7. *London Plane, Carshalton Ponds, Surrey*
The Carshalton Beauty. Dominates Festival Walk beside the River Wandle; not quite the tallest or biggest, it is among the best allround, 135ft × 23ft and growing fast.

8. *Silver Lime, Tortworth, Gloucestershire*
The other side of the church from the famous old sweet chestnut, and unnoticed by Elwes, Henry, or other pilgrims to that tree. It stands 111ft × 14ft 8in, by far the biggest in both dimensions, superbly handsome foliage, well-sited and growing very fast.

9. *Common Alder, Sandling Park, Kent*
Hard to believe it is an alder, not a prime oak. All but 100ft tall and 13ft 3in round the smooth, grey cylindric bole. It has long been the finest of its species.

10. *Trident Maple, Norham Gardens, Oxford*
I found this in about 1958 and never really believed that I was right; that a very rare and special Chinese tree should grow in suburban surroundings of a little front garden, to twice the majesty of any in the prime tree collections, must be explained by faulty identification in my inexperienced days. Not at all. I return to worship it at intervals and none is yet known anywhere in its class.

My list omits many trees which deserve your attention, like the original ginkgo at Kew, the incredible London plane at Lydney Park, the oriental plane at Rycote and all those superb, towering 150ft tall giant sequoias and 180ft tall grand firs in Scotland. My selection aims to provide variety of location and species, and I wanted to include trees which are, on the whole, accessible to the general public.

5. FLOWER-WATCH

The natural plant collections of Britain and other countries in the northern hemisphere have had to survive the ravages of several ice ages and are thus relatively young and sparse. The last great ice age, some 10,800 years ago, destroyed our vegetation, including all our trees. Two thousand years later, by the time Britain had become an island separated from Europe, the climate had warmed up enough for about 35 species of trees to become re-established, the seeds having been brought in from Europe and elsewhere by birds, the sea and the wind.

It is a great compliment to British horticulturists that we have succeeded in creating a unique collection of imported flora, but if you wish to see these plants in the real glory of their natural state, you have to travel to the places that recent ice ages failed to reach.

The most prolific natural flower garden on Earth is to be found near the southernmost tip of Africa – a spectacular peninsula which was rightly named 'the fairest Cape in the circumference of the Globe', by Sir Walter Raleigh when he rounded this spectacular peninsula in the *Golden Hind* on 18 June, 1580.

The greater part of the vegetation here is of a type called 'sclerophyll'. The plants have tough, rigid or leathery leaves which are resistant to drought and water loss. Locally it's called the 'Fynbos' – the fine bush – which understates a flora that is unrivalled on Earth.

TONY HALL:
Cape Floristic Kingdom

Today a squadron of plant-watchers study the kingdom from their base at the University of Cape Town. We joined one of them, Dr Tony Hall, for a tour of an area which must make a very strong claim to being the Garden of Eden.

As we walked down a valley, with the Indian Ocean behind us and the Atlantic filling the horizon in front, Tony told us that in this short stretch of about a mile were almost certainly some 1500 species of flora. (In Scotland you could search all day and only find nine or ten species.) 'It actually has more species of wild flower than the whole of the United Kingdom

Opposite: *Garden 'standards' originate in profusion around the Cape including Salvias, Agapanthus, Pelargoniums, Gladioli and Watsonias – and a profusion of heathers; here competing with the more exotic Mimetes.*

Above: *Tony Hall: high priest of the Cape Floristic Kingdom. This spectacular peninsula at the tip of Africa was Walter Raleigh's 'Fairest Cape in the circumference of the Globe.'*

Below: *Cape Floristic Kingdom, South Africa.*

TONY HALL
Cape Floristic Kingdom

Below: *Amid the innocents, beautiful carnivorous plants like this drosera lure insects to their doom in a sticky flypaper of vegetable matter. Most carnivorous plants like damp conditions and the family is highly threatened by dam-building and drainage works.*

put together. (We have some 1300 in total.) They grow in an enormous range of habitats and have been here for an astonishing length of time – at least 80 million years – without any major disturbances. It is widely believed that this may have been one of the original points of evolution of the southern hemisphere flora.'

Everywhere one sees familiar plants which may have originated here long ago and have since become common in cultivated gardens throughout the world.' Geraniums of several different types sprout out of the poor, sandy soil and there is a wealth of *Salvias, Agapanthus, Pelargoniums, Gladioli* and *Watsonias*. An incredible 600 species of heather colour the stony hillsides.

The Cape Floristic Kingdom is believed to contain the richest abundance of species of flora in the world; it is certainly the most concentrated. There are known to be 8000 species in an area as small as Wales. Therefore it is hardly surprising that the earliest botanical expeditions resembled looting raids.

'Enormous quantities of specimens were sent back to Europe, even to the great Linnaeus himself, who was staggered by their abundance. A large area of his famous herbarium is taken up with Cape plants,' Tony told us.

Paradise lost?

The future of this quite remarkable garden is, however, in serious doubt. One of the reasons the plants have done so well is due to the Cape's benign climate. The peninsula lies just far enough south to catch the cyclones and depressions that bring cool, moist winds and winter rainfall. In summer, southern hemisphere high pressure belts bring warm, dry weather.

These are also ideal conditions for humans. The population in the metropolitan region has increased by 33 per cent in just ten years, and the country's population as a whole is growing by 60,000 a month. For the Cape Floristic Kingdom such population growth is very serious because it means that more and more of it is appropriated for agricultural purposes. It is an important food-producing region, and is expected to provide larger and larger quantities of food for the ever-expanding population.

'We had an awful shock when we studied recent satellite pictures and discovered that in fact only 39 per cent of the Floral Kingdom is left,' Tony told us. 'The rest has been infested by alien vegetation and human development.' By African standards this is hardly serious, but Tony and other nature-watchers refuse to regard the Kingdom as anything other than unique.

'We know that a total of 28 species of plants have become extinct over the last century. (These include unique species like the protea *Aulax pinifolia*, used for firewood.) That may not seem too large a number in a world that's losing species and resources every second, but in fact it represents 90 per cent of the total plant extinctions in southern Africa.'

What is serious is that a number of the plants found in the Cape Floristic Kingdom are endemic (they exist nowhere else). One of these is *Leucadendron macowani*, an unspectacular leucadendron resembling a cross between a stunted pine tree and an Australian gum. The family to which it belongs occurs in all the southern continents, and its roots go back to the earliest flowering plants, 100 million years ago. The poor little *Leucadendron macowani*, however, is now cornered.

Above: *The Cape is famous throughout the world for its proteas of which the most spectacular is the giant variety. Individual blooms sell for several pounds in European flower markets, encouraging a unique Cape criminal activity – plant poaching.*

TONY HALL
Cape Floristic Kingdom

There are just 180 left and they mostly live on one small patch of marsh, which we visited. One accidental fire could eliminate this plant from the face of the earth.

Other Cape plants and plant families are facing the same situation. A few miles away, where there are plans to build a large dam to supply the city of Cape Town with electricity, we were shown the endangered Marsh Shrublet, *Spatalla prolifera*, who's family exists only in the Cape. Nearby, almost invisible underfoot, there is a *Roridula gorgonias*, one of the insect trapping plants, which is also found nowhere else, and is regarded as critically rare.

Above: *The beauty of Table Mountain, South Africa, rivalled only by the unique plants like this clump of yellow arctotheca, growing all around the mountain.*

Paradise regained?

Tony Hall and the other custodians of the Kingdom insist that an area which is the root stock of some of the world's most famous and rare plants requires extra special conservation. They have evolved a three-pronged plan of action as follows:

1. The Kingdom's botanists are tackling the formidable task of cataloguing this vast flora, with the aid of computers. A decade of research has revealed some interesting and alarming facts. For example, 28 species have recently become extinct. There are 96 endangered species, 125 that are vulnerable, 336 critically rare and another 500 or so whose status he is unhappy about. These include the Golden Gladiolus *Gladiolus aureus*, the heathers *Erica laakii* and *E. annectens*, and a very rare iris *Witsenia mauria*.

His detailed plant counts have also revealed another danger which botanists have long suspected: we are probably destroying plants before we even know of their existence. If it is true in the closely-monitored Kingdom it must certainly be true is less supervised habitats such as the tropical rain forests.

Tony Hall regards this list as alarming, and he is using it as ammunition, with considerable success, to call for the establishment of 19 major plant sanctuaries; this would protect the Kingdom from end to end. Tony and his colleagues have already set up ten sanctuaries – possibly a record for plant protection on this planet.

2. They are waging a war on 'aliens' (highly resistant foreign plants), in particular Australian wattles and spiky *Hakeas*, that have invaded the Kingdom in strength to the detriment of the native flora. Tony explained: 'One shoot of wattle can carry as many as 15 seed pods and there are ten seeds in each pod. So this little branch will create 150 plants that may live 150 years. In certain places the situation is quite hopeless; it is a very serious problem.' Most of the aliens produce a juicy fruit that is attractive to birds, and the green groves of wattle attract picnickers who also help to spread the invasion. A friend of Tony Hall's once examined the detritus from a car-wash and found enough seeds to produce 18,000 plants.

The Cape possesses none of the insects that control the 'alien' plants in their home countries; indeed, their prodigious seed production evolved to cope with pressure from the predators they left behind. So Tony Hall and friends have let the world's most effective predator – Man – loose on the aliens; it is the only pro-conservation slash-and-burn operation we have witnessed. Every weekend, specially recruited teams of active retired people clamber up the steep

TONY HALL
Cape Floristic Kingdom

Flowering Plants

The angiosperms, or flowering plants as they are more commonly known, are plants whose seeds are protected by an ovary. They include all our familiar garden plants and wild flowers, bulbs, palms and grasses.

Their origin, to quote Charles Darwin, has always been 'an abominable mystery'. Darwin's despair was based on the fact that then (more than a century ago) and now, there is a complete lack of fossil remains to provide plausible connecting links between the angiosperms and any other group of vascular plants. Theories about their origin are therefore not much more than enlightened guesses. At the moment three different hypotheses are in vogue:

1. The 'classical' theory suggests that angiosperms evolved sometime during the Mesozoic era (65–225 million years years ago), in equable tropical forests, from ancestors similar to the seed ferns found in fossil form in the late Paleozoic (225–570 million years ago). They were tree-like with large flowers resembling present day magnolias.

2. The second 'contemporary' theory, regards the angiosperms as a single cohesive group with all the various orders and families radiating from a common ancestor. There is no one family directly related to this common ancestor: several modern families are equally closely related to it. The links are supposed to be extinct.

3. The third and perhaps most advanced theory (in fact a group of theories) holds that the angiosperms have a multiplicity of origins and that the *Gnetales* (gymnosperms) hold clues to the origins of many groups of flowering plants.

There is more agreement on the general shape, size and habitat of the original flowering plants. It is thought that they evolved not in the tropical rain forests of the Cretaceous period (65–120 million years ago), which has

continued

TONY HALL
Cape Floristic Kingdom

continued

a rich fossil record, but possibly as far back as the Permian (225–280 million years ago) in upland regions with an equable, tropical or subtropical climate and seasonal drought. They were thick-stemmed, large-leafed trees bearing large fruit. Adapting to the rigours of a drought cycle is generally regarded as one of the factors which promoted the rapid evolution of the primitive angiosperms, particularly the need to protect or hide their genetic legacy, seed. Drought periods are also periods of calm weather and, in that they follow a 'rainy' season, are times of maximum insect activity. Both these factors are thought to have encouraged the primitive angiosperms to evolve from wind to insect pollination.

slopes of the Kingdom armed with saws, spades and machetes to cut the native plants free from the aliens. They all seem to have a good time and share an understandable sense of achievement. Finding a glorious plant like the giant protea (p. 81) in the middle of a dense thicket of wattle must be as exciting as finding a lost city in the jungle.

3. The Kingdom's official plant-watchers have thrown their weight behind the establishment of a lucrative and well-run plant industry. Cape plants bloom to their full glory just when the northern hemisphere is gloomy and cold and most in need of a few bright flowers. They travel and keep well. A single bloom of some of the more exotic proteas can fetch a high price in London, Paris and Hamburg.

Unfortunately this endeavour has also encouraged plant poachers; some of them have been known to clear an area of plants worth US $50,000 on the European markets. To stem the poaching Tony Hall believes the Cape public need to be more appreciative of the fact that they live among some of the finest flora in the world, which has great commercial value if properly protected.

Enlightened and aggressive though the Cape conservation movement may be, it still has to tackle the difficult question of human population size in the area. 'When are we going to relate our total population numbers to the needs of the environment?' Tony asked. 'That is the most crucial question we have to face in the next decade.'

In his opinion it is also a question that taxes our morality: 'We have a responsibility to other species that share this planet with us,' he emphasized. 'They have a right to a future. They should be able to stay as they are, to expand and flourish or even to become extinct: but on their own, in their own time, naturally!'

This is an idealistic concept and for the time being plants may have to be protected in special sanctuaries – places like the Royal Botanic Gardens at Kew, England.

JOHN SIMMONS:
Kew Zoo

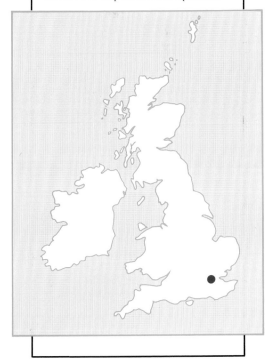

Human attitudes towards plants are conditioned by the fact that they are organisms from a non-human dimension. We have few inhibitions about destroying these complex life forms (and none at all about eating them) and it is only in the last decade or so that threatened 'habitats' have been accorded the same importance as threatened animals by the conservation movement.

One small group of nature-watchers may be excluded from this condemnation, the botanists. Long before facilities were provided for animal rescues, endangered plants were being 'captive bred' and carefully husbanded in botanic gardens, of which the most famous is the Royal Botanic Gardens, Kew.

This is a most unlikely setting for a sanctuary of any kind. On the outskirts of London, directly under the flightpath to Heathrow Airport, it has poor, sandy soil and adjoins the River Thames which for most of Kew's 200-year history has been highly polluted.

Kew consists mainly of two estates bought by George II, his wife Queen Caroline, and their son, Frederick in 1720 and 1730. When Frederick died his widow, Augusta, started a botanic garden on the two properties and erected a number of famous buildings, including the Orangery, Pagoda and the Ruined Arch, which still stand.

Kew's plant collection became internationally famous a decade later when Sir Joseph Banks took charge and sent young men all over the rapidly expanding British Empire in search of plants. There was a period of decline after Banks's tenure, but this was reversed by the appointment of Sir William Hooker in 1840: he arranged for the grounds to be landscaped (by W. A. Nesfield) into vistas, walks and avenues, and the famous Palm and Temperate Houses were built.

Sir William was succeeded by his son, Joseph, who was already a famous botanist and explorer in such remote areas as Nepal and Sikkim. Joseph Hooker established Kew as a centre of scientific research, in particular for the formal classification of plants and their distribution.

Another period of decline descended on Kew between the wars until, in the 1970s, two significant appointments were made; Professor Heslop Harrison developed the 'seed bank' which gave Kew a new role

JOHN SIMMONS
Royal Botanic Gardens Kew

Above: *John Simmons, with the famous Kew cycad (p. 88).*

Below: *The map shows Kew's position.*

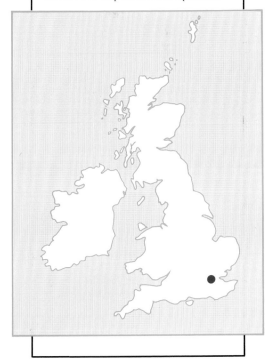

JOHN SIMMONS
Royal Botanic Gardens Kew

in plant conservation, and a lively young gardener, John Simmons, was made Curator of the Living Collections, which compliment the scientific and conservation work of the Herbarium, Laboratory and Kew's country extension at Wakehurst. John Simmons and the other gardeners and botanists at Kew have restored its reputation as the world's leading botanical garden, but, equally important, they have enhanced the public appeal of the place.

Tens of thousands of visitors now come to Kew from all over the world to enjoy the variety of mature, well-positioned trees, and the wealth of flowers and shrubs.

This delicate balance of aesthetic pleasure and science is Kew's greatest achievement, typified perhaps by the walled Herbaceous Grounds where the public may wander amidst a wealth of colour, while

Right: *The Palm House at Kew where 'living fossil' plants thrive, and hundreds of years of sanctuary have been offered to rare plants from the furthest corners of the world; indeed some of the earliest-collected species are now extinct in the wild, leaving Kew with the only examples left.*

botanists (identified by their notebooks) stand and stare. Such tidy flowerbeds are something of a threatened habitat in our gardens because of the constant work they demand, but John Simmons has always been a staunch defender of the herbaceous beds at Kew.

'They are a historic treasury in their own right,' he believes. 'The Herbaceous Grounds, as they are called, were laid out in 1840. Today we keep between 2500 and 3000 species there as a living reference collection for scientific use. Overall there are probably upwards of 18,000 herbaceous plants tucked away in various corners of the gardens: you can never be entirely sure because it increases daily with the enthusiasm of our staff.'

Simmons uses the Herbaceous Grounds when he needs to convince a lay visitor of the value of Latin names for the identification of plants. He took us to see a collection of poppies from all over the world and pointed out that in spite of their widely diverse habitats, many looked remarkably similar.

'You have to have a language of taxonomy that can be used universally, and Latin is as good as any,' John suggested. 'If you regard plant names as the surname and christian name of the plant, you soon get the hang of it. The surname identifies the family while the christian name identifies the individual. If you were a plant you would be Pettifer Julian,' he told Julian.

Looking round, he quickly found examples to illustrate the importance of a detailed naming system. 'Over there we have a maple. Its common name is sycamore. In Israel, however, a sycamore is a fig tree, and in America it's a plane tree. Those chestnuts over there – a sweet chestnut and a horse chestnut – are not related at all. If you stuck to their Latin names you would know that.' (They are *Castanea sativa* and *Aesculus hippocastanum* respectively.)

Simmons also regards the Herbaceous Grounds, backed by Kew's huge specimen collections, as important tools for the conservation of plants and animals; indeed of whole habitats. 'We have had people studying great apes who want to know about their food plants. From the live and reference collections we have here we can identify the samples they send us, name them and explain their importance.

'You can't have a tiger without a forest, so to speak, but we have yet to properly appreciate the absolute

JOHN SIMMONS
Royal Botanic Gardens Kew

JOHN SIMMONS
Royal Botanic Gardens Kew

interdependency of plants and animals. We were consulted at Kew about some Hawaiian plants that had stopped setting seed and were rapidly heading for extinction. From our very old records we revealed that these plants were pollinated by a certain bird species – which was already extinct on the islands.'

Some of the plants alive at Kew today have an importance that goes far beyond their reference use. In the recently-restored Temperate House (surely the most beautiful glasshouse in the world) John proudly

Above: *Kew's Temperate House, a work of art in its own right and probably the most famous greenhouse in the world. That such a structure has survived the expansion of London (Kew is now under the main flight path to London Airport) is an indication of the love city dwellers have for these beautiful gardens that can be reached by underground train from the centre of London.*

showed us his collection of 'living dinosaurs', the cycads. If you discount the tree ferns, these forerunners of the conifers are regarded as the oldest trees in existence. They were already established in the Jurassic period (160 million years ago) when the dinosaurs trod the earth.

One Kew cycad in particular is totally unique. It was collected from South Africa at the turn of the century and its species became extinct in the wild in 1918. The Kew cycad is a male, and no female has ever been found.

This lonely living fossil underlines Kew's importance as a sanctuary; as a place where a plant can be totally protected while Kew scientists explore new methods of reproducing these incredibly rare organisms.

Ever the practical gardener, John pointed out that the South African cycad 'is rather like a carrot – you should be able to cut the stem off and root it in a bed of sand. But we don't have the courage to do that with this old fellow who is absolutely the last of his line. We're trying tissue culture from small pieces of leaf, but so far we haven't got it through to a full plant.'

This miracle-work, or cloning, goes on in a part of the laboratories called the Micropropagation Unit. A room of jars and test tubes full of green things in grey jelly, its real significance is not immediately obvious but here scientists are evolving techniques that should mean no plant need ever become extinct.

'The theory is that you can grow any plant from a single cell, and it should be exactly identical to the parent plant. In practice it is much more difficult than that. Some plants, like the strawberry runner, have a natural propensity for propagation; others have a totally different anatomy and structure and they don't easily reproduce their parts. Frankly, it is not very well understood but we are gradually learning how to reproduce more and more plants.

'We work here in almost aseptic conditions. Tiny scraps of plant are rooted in a medium (nutrient jelly) and we manipulate them with hormones. Some hormones make roots grow, others work on the shoots.'

The advantages to horticulture of cloning are obvious. If you raise natural seedlings, each seedling is genetically different. Cloning allows plants selected for resistance, high yield etc., to be produced in almost infinite numbers. Kew's Micropropagation Unit is more concerned with the cloning of endangered plants. John proudly displayed half a dozen jars containing sprigs of cloned endangered plants from as many countries. In most cases, these jars contained many more plants than exist in the wild.

'You can easily lose a plant altogether when its numbers become dangerously reduced in the wild,' John pointed out. 'We do much better here. Our purpose is to reintroduce these to the wild, sometimes finding a completely new home if, as is so often the case, their endangered status comes from the destruction of habitat.'

The Kew unit has, in fact, caused certain plants which were regarded as notoriously difficult to propagate and become common. Almost all the Venus flytraps sold commercially in Britain are descended from stock first cloned at Kew. 'It is a valuable stock in more ways than one,' John pointed out. 'Firstly we

JOHN SIMMONS
Royal Botanic Gardens Kew

JOHN SIMMONS
Royal Botanic Gardens Kew

Above: *When the nuts of the Coco de Mer first reached Europe they sold for thousands of florins – as an aphrodisiac. They have always been associated with fertility, perhaps for obvious reasons. When General Gordon discovered them growing on the Seychelles he decided he had found the Garden of Eden.*

have what amounts to an infinite supply of Venus flytraps – as many as you want, for ever and ever! Secondly, it means that plants of this kind need never be taken from the wild.'

Plants bizarre

For much of the public, the great appeal of Kew is the exotic nature of the collection. More modern botanical gardens tend to focus on native species, but Kew, because of its long history, is a vast job-lot from all over the world, an unmatched collector's box that is a legacy from our days as an Empire.

In the Palm House, another masterpiece of Victorian glasswork and wrought iron, John showed us the largest nuts in nature and revealed that an eminent Victorian, General Gordon, once informed the world that they were the fruit of a tree from the original Garden of Eden.

This nut is the double Coco de Mer (see left) which, for obvious reasons, has always been associated with fertility and once sold in Europe for up to 10,000 florins (£1000) as an aphrodisiac. The nuts float and the first specimens to reach Europe (found off the Maldive Islands) were thought to emanate from strange trees which grew on the floor of the ocean. The Seychelles, where these 'coconuts of the sea' actually grow, had yet to be discovered. When they were, General Gordon decided the shape of the nuts was so singular that the Seychelles had to be the original Garden of Eden, and he published a learned paper in support of his theory.

'The Coco de Mer is in fact a very ancient tree in a very ancient land,' John acknowledged. 'Its method of propagation is also quite remarkable and it took us years to get one going. The huge nut is a food store on which the seedling feeds for about five years! Unlike other nuts which project a shoot or root, this thing pushes out an umbilical cord-like growth at least three feet long at the end of which the seed germinates.'

The Coco de Mer is not the only plant at Kew with a remarkable growth rate: one such has its own heated house and a public following which has remained faithful since the plant, the Giant Water Lily, *Victoria amazonica*, came to Kew a century ago. A new plant is raised each year from seed and planted in a huge submerged box in the Tropical Water Lily House, built especially for the plant in 1852. Growth is almost visible as the huge spiked pads push outwards at a rate of almost a square inch a minute until they reach nearly six feet and can support the weight of a small child.

Almost everywhere you look when at Kew, you will see plants with a story behind them. We even came across two large palms that appeared to be dying, a very unusual occurrence here. John informed us sadly that even botanic gardens cannot escape politics: he believed these two palms, which were given to Kew by members of the Royal Family, had most probably been deliberately poisoned. A similar fate overtook a tree that had been planted at Kew by the Emperor of Japan a few years ago. Fortunately such vandalism is extremely rare.

We cannot pretend to have presented anything more than a glimpse of this remarkable garden or indeed of the skills and knowledge of its head gardener, John Simmons. We did, however, ask him whether it was true that certain people had 'green fingers' and if so, how did one acquire them?

'The difference is between gardeners who think people rather than thinking plants,' he smiled. 'You've got to think in terms of what the plant wants, not what you want. You can tell if it wants more light or more moisture simply by looking at it. So that's it really, just think plant.'

JOHN SIMMONS
Royal Botanic Gardens Kew

Below: *Kew's crowd-puller, the trifid-like* Victoria amazonica *water lily. John Simmons and his gardeners are hard-put to keep up with its extraordinary growth rate: from a new seedling to these six-feet-wide pads every year. Growth is actually visible – a square inch a minute!*

6. INSECT, SPIDER AND SNAKE-WATCH

About 350 million years ago the red quartz deserts of the preceding Devonian epochs were replaced by the 'dripping wet, mist-drenched, luxuriant rain forests' (Wendt) of an era known as the Carboniferous. The surface of the Earth became a moist, fertile loam. The embryonic plants exploded in this virgin garden and in the next 50 million years virtually the entire planetary surface became one vast tropical jungle. Experts are constantly debating the causes of this, but no theory has been settled on yet.

In retrospect, the Carboniferous is not only the dawning of a new era for plant-watchers, but also Day One for the entomologists (insect-watchers).

In this giant forest, the largest phylum in the animal kingdom arose; the arthropods. The spiders, millipedes, and a kaleidoscope of insect life abounded. Swamp forests afforded them the ideal conditions for life, as they continue to do today. In the Carboniferous swamp forests there were dragonflies with wingspans of six feet. The insects still command the record for sheer diversity of body form among living creatures on Earth and they are amongst the most successful of all organisms.

Here we shall focus on two aspects of the world of insects: (1) their capacity for social organization such as is found in honeybees and (2) their extraordinary and powerful means for defending themselves against enemies using chemical weapons – toxins that can also affect man.

BROTHER ADAM:
A life with bees at Buckfast Abbey, Devon

We were fortunate enough to be able to embark on a bee pilgrimage with an 84-year-old German who is widely regarded as the world's greatest keeper and breeder of bees: Brother Adam, a Benedictine monk, who has lived and worked at Buckfast Abbey in Devon since 1911.

After three years at Buckfast, the young Brother Adam fell ill and the Abbot took a fateful decision about his future. 'While considering what to do with me,' Brother Adam smiled 'the Abbot set me to bee-keeping, thinking that perhaps an interest in bees

Opposite: Assasin bugs use camphor scraped from plants as a natural insect repellant to protect their eggs.

Above: Brother Adam of Buckfast Abbey, Devon, saviour of the British bee; we accompanied him on a trip to Greece in search of pristine bees from the Golden Ages.

Below: Buckfast Abbey (Britain) and Mount Athos (Greece).

would help me find my way. Of course he was right: there is something fascinating about bees. Several poets, and even Aristotle wrote a great deal about bees; they have fascinated intellectuals since the beginning of time.'

The Abbot's decision was also a very fortunate one for British bees. In 1915 these islands suffered what became known as the 'Isle of Wight disease'. This bee-disease (called acarine disease and caused by a parasitic mite, *Acarapis woodi*) spread to the mainland from the Isle of Wight and in 12 years exterminated 80 per cent of the bee colonies in the UK. In one year the 46 thriving colonies at Buckfast were reduced to 16. Most of the survivors had originally been imported from abroad, which led Brother Adam and the other Buckfast beekeepers to the conclusion that amongst these were strains of bee that were resistant to the Isle of Wight disease.

To help re-establish Britain's honeybee colonies, Brother Adam set up the world's first bee-breeding station on the remote slopes of Dartmoor; he began a selective breeding programme of queen bees – the reproductive founts of bee colonies.

His success has made the Buckfast Bee, and himself, a legend in the world of beekeeping. To appreciate the science, and the art of bee-breeding, one must understand something of the intricate social and sexual life of the beehive.

A honeybee's life
The 50,000 bees of a natural colony work together as one unified community; each bee performs its own specific task. The majority are sterile female worker bees; they travail as food gatherers, builders, cleaners and nursery maids, they are a powerful and adaptive domestic labour force in the hive. In addition there are about 500 male drones who do no work. In time just a few of them will mate with the only fertile female in the hive, the solitary queen.

The young queen, when perhaps only a week old, takes off on her nuptial flight. Only the strongest of the male drone bees can keep up with her, and will succeed in mating with her. The act literally kills the drones but their supply of sperm will be stored in a sperm-reservoir, and lasts the queen for the rest of her five years.

The queen may lay up to 3000 eggs in a day, constantly replacing the workers of the hive who live for only four to six weeks. All the eggs look alike, but those laid and then fertilized will become sterile female workers. Those laid without being fertilized become male drones.

The queen herself emerges from a selected worker egg which has been fed on special rich 'royal jelly' (secreted by female workers) throughout its development. She is the all-powerful ruler, controlling the day-to-day behaviour of her subjects by using chemical signals (pheromones). Beekeepers call these chemicals 'Queen's Substance' which the workers lick off her. Pheromones appear to be essential for the maintenance of the morale and cohesion of the colony.

The fact that all the queen's subjects are also her own progeny, gives each hive an individual character, as Brother Adam confirmed. 'In every hive there are differences, although some are only slight. Some hives for example produce significantly more honey than others: if I change the queen, the characteristics of the new queen will be reflected throughout the colony. It is one of the most fascinating aspects of beekeeping. No two colonies are exactly alike.'

Breeding better bees

Detailed observation of these individual hive characteristics has enabled Brother Adam to selectively breed features that are attractive to beekeepers: 'Just

Royal Jelly
Three days after laying, nurse bees surround the eggs with 'bee milk', also known as 'brood food' (or royal jelly), ready for the emergence of the larvae.

This milk is a highly nutritious fluid secreted by the hypopharyngeal glands in the heads of the nurse bees, and all larvae receive this food for the first two days of their life.

From then on only queen larvae are fed generous quantities of this rich diet, hence the term 'royal jelly'. Worker and drone larvae are reverted to a diluted diet of royal jelly, pollen and nectar in restricted quantities.

Below: *In a normal bee colony, only the first queen to emerge lives, and she kills all others. Brother Adam ensures his elite queens survive by hand-rearing them in isolation, then establishing them in miniature colonies in which there is no competition. Here bee number 59 is the queen bee.*

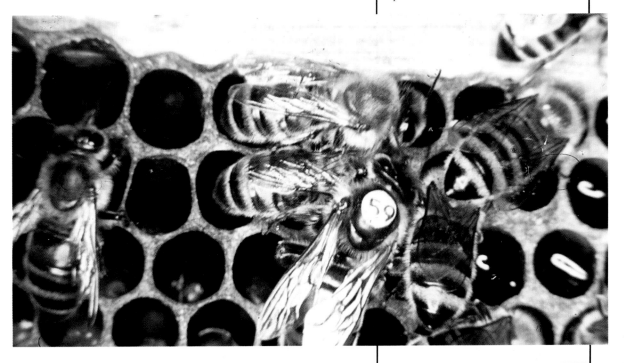

BROTHER ADAM
Honeybees

before the war I developed a bee from a cross between the French black bee (which is very bad-tempered) and our own bee, with the result that we had a strain which virtually didn't sting at all.' Brother Adam hoped he would produce a bee with the good nature of the English bee and the prolific, hard-working characteristics of the French bee, and he was very successful. 'You could kick the hive and disturb it as much as you liked (bees kept for honey production are disturbed quite a lot) but unless you squashed one of the bees you wouldn't get stung. Unfortunately that strain was also highly susceptible to disease so we had to give it up.'

Brother Adam's idea of the perfect honeybee is one that is extremely good-tempered, prolific in honeycomb production, resistant to disease and one that does not leave the hive periodically to 'swarm', as this entails a great deal of work for the beekeeper.

So how does he go about producing better bees? As the character of the bee colony depends on the character of its queen, it is the rearing of the queens that is important.

Under natural conditions the population of a healthy colony grows in the spring. By adding living space, more and more bees are hatched to fill it. Pressure is intentionally built up, forcing the bees to produce new queens which then leave the hive in a swarm with some of the workers; by so doing they divide the colony.

Instead of letting the bees breed their own queens, Brother Adam cleverly introduces hand-reared queen larvae from selective breeding that he hopes will produce the qualities he wants. They are given their own man-made 'queen cup' and fed lavishly on the vital royal jelly. Expanding hives readily accept these 'Alpha Plus' queens if they are introduced at exactly the right time.

Problems, however, can arise in this rearing process when the queens emerge. Left to their own devices, the first to bite her way out will immediately attack adjoining cells. Queen bees will not tolerate competition and they will fight to the death with other queens.

Brother Adam has overcome this difficulty by rearing the queens in individual cells. Later they are transplanted to miniature colonies where there is no competition. It is from these hives that the specially reared virgin queens will fly and mate, but again only under Brother Adam's benign and watchful eye and only with his hand-picked drones.

Brother Adam took his bees to remote Dartmoor because there the queens are beyond the range of other possible suitors who could do damage to his careful control of the genetic line. The chosen characteristics are thus passed on by the mated queen to the entire colony.

Has he in fact produced the perfect honeybee? He smiled gently and said, 'I cannot even hope to produce the perfect bee. But the Buckfast Bee does better than any existing ones.'

He is, however, haunted by a problem that troubles all genetic breeding programmes: the hazards of inbreeding caused by continually mating bees that are genetically related to one another (like members of one family). As a result, Brother Adam has travelled the world, looking for new and suitable bees. In the past 30 years he has searched through Spain, Portugal, Turkey, Yugoslavia, Israel and the deserts of the Arab world, a royal matchmaker in search of the perfect princess for his strapping yeoman drones. No beekeeper has ever travelled so far or searched so diligently. Indeed, Brother Adam's search for the golden hive may be as unending as Jason's for the Golden Fleece for, at the age of 84, he announced that he was off again to the home of the Argonauts and invited us to join him.

A bee pilgrimage

It was a magical expedition to what must be one of the oldest centres of human worship of the bee, the holy island of Athos, with its ancient monastery of bee-keeping monks. The island-peninsula of Mount Athos is almost cut off from the mainland and the Macedonian bee was known to exist there until 30 years ago.

Four thousand years ago the great civilization of the Minoans held the bee sacred, and Brother Adam considered Mount Athos the only place where he might find an original Macedonian bee.

'No foreign bees have ever moved to Athos,' said Brother Adam hopefully as we moved across a silver sea in the direction of some buildings that looked as if they had been built in the Golden Age and had not been repaired since. 'All over the rest of Macedonia, bees have been brought in from other parts of Greece, with the result that the original bee has disappeared.'

Our landing was a step back in time. Almost the only inhabitants now are Greek and Russian Orthodox monks. Virtually the only buildings are the huge monasteries which look almost as though they are

Bee-keeping: Modern hives consist of a box (brood chamber) open top and bottom, set on a floorboard, crowned with another board (the crown board) capped with a waterproof lid, or roof. The hive is extended by adding further boxes (supers). Wooden frames hang in these boxes holding the combs.

We recommend you consult a local bee-keeper before choosing a hive. Six designs are available in Britain (W.B.X., National, Modified Commercial, Langstroth and Modified Dadant), but the so-called 'beginners hive', the National, may not be the best because it is built to accommodate an extinct bee! (Or so we are advised.)

'Standard' British brood frames (14in by $8\frac{1}{2}$in) were fixed 100 years ago by the British Bee-Keepers' Association when the English Black still ruled the roost (or hive). Its replacements, including Brother Adam's Buckfast, are much more prolific and, ideally, need a big hive. So consult a local expert. You will probably be advised to acquire a Modified Dadant with a single large brood chamber containing 11 large frames.

For further information:
British Bee-Keepers' Association
(For local bee-keeping associations) Secretary: M. H. F. Coward, High Trees, Dean Lane, Merstham, Surrey RH1 3AH.

British Bee Journal
46 Queen Street, Geddington, near Kettering, Northants NN14 1 AZ.

BROTHER ADAM
Honeybees

in their original condition. The community was established on Athos before William the Conqueror came to England in 1066 and it is the world's longest-lasting democracy. But most remarkable of all, for the past 800 years no female has ever been allowed by the monks to set foot on the place. Not only no human female, but no female domestic animals – no cows, no ewes, no nanny goats, no bitches, no she-cats.

Fortunately for Brother Adam the rule does not include female invertebrate animals. There are queen bees on the Holy Mountain and in spite of his advanced years, Brother Adam was soon striding up the steep, stony paths in search of his princess.

Right: A strange monastic sect at their devotions? No, Brother Adam and the beekeepers of Athos defensively garbed against the ferocious local bee.

A number of likely candidates were selected and carefully packed for shipment back to Devon. These will be carefully crossbred for a number of years to see whether they can contribute to Brother Adam's dream of the perfect honeybee. At the very least, if the breeding plan fails he will have eliminated one possible solution to his problem. And that may be sufficient for Brother Adam, who learnt long ago that patience is not just a suitable virtue for a monk, but also a prerequisite to successful beekeeping. It is probably one of the most therapeutic forms of nature-watching, and it has a bonus: delicious honey to spread on your toast, while you ponder the wonders of your hive and its inhabitants as they go about the business of improving your fruit trees and bushes, and make gentle music all over your garden.

Post script
Brother Adam informed us in August 1984 that the Athos queens had been used for crossbreeding on an extensive scale, and had given 'promise of great possibilities'. He added with typical caution: 'It will take another year before I am able to make a conclusive evaluation.'

TOM EISNER:
Chemical defences of insects

Food for thought
One of the more bizarre features of the eating habits of arthropods is that sometimes they consume substances which are then converted into poisons or traps, for defence.

Dr Tom Eisner is Professor of Biology at Cornell University in the USA. Through years of painstaking study he has become the world's greatest expert on the methods that insects use to stay alive. These survival mechanisms, or as Eisner sees them, defence mechanisms, have probably been responsible for the essential structuring of the insect world.

During our visit to Cornell, Tom took us on a field trip and then to his laboratory to demonstrate the quite extraordinary mechanisms which these tiny creatures have developed to stay 'ahead' in the teeming jungle of life.

Termites
He began with the very common social termite (*Nasutitermes exitiosus*), a soft-bodied creature which, without a system of defence, would be vulnerable to almost everything in the insect jungle, including creatures as tiny as ants. The wariness of termites is so pronounced that they will defend themselves against anything that moves. When Eisner created an artificial ant from a spinning sliver of metal and put it in their path, he noticed that the soldier termites turned

Above: *Tom Eisner: he uses the canyons and ravines around his home in New York State to find examples of the secret weaponry of the insect world.*

Below: *Cornell University, New York.*

TOM EISNER
Insects' defences

Termite Toxins

Termites (Order **Isoptera**) are known throughout the world as 'white ants'. In fact termites are no more closely related to ants than mice are to men. Admittedly they look and act like ants at first glance and they have a social life and organization which closely resembles the true ants. This, in itself, is a quite extraordinary example of parallel development, as remarkable as say our finding that kangaroos lived exactly like human beings when we discovered Australia.

Two thousand termite species are known to exist (a count which has doubled in the last years), mostly in the tropics, and they are the most primitive of social insects. They are believed to descend from ancestral stock resembling present-day cockroaches.

It is almost easier to consider a colony of termites as a single organism rather than a group of individuals. The core of this rigidly-organized mass is the king and queen whose function is to breed; their 'arms' are the workers, and the whole is guarded by soldiers. The colony is held together by chemical signals that are either smelled or tasted.

The colony is defended by an arsenal of chemical arms employed by the soldiers, and the weapons synthesized by termites are quite unique. Every soldier has a 'tank' of these chemicals built into its body. When attacked, most commonly by ants, termite soldiers attempt to puncture the cuticle of the enemy, then coat it with a paraffin-like oil which interferes with the coagulation of the victim's hemolymph (insect blood) and inhibits resclerotization (natural repair of surface tissue). No other animal or, for that matter, plant, is known to synthesize these compounds.

continued

Right: *Close-up of the head of a nasute termite. Note that the snout is producing a large droplet of 'glue' to be used to incapacitate an opponent.*

to face the threat and did something which immediately caused other soldiers to come scurrying to the attack. The reinforcements then also squared up to the enemy and stood their ground.

Soon Eisner observed that his artificial ant, turned by a magnet, was slowing down, until finally it stopped. Only when he turned a powerful magnifying glass on the soldier termites was he able to detect the

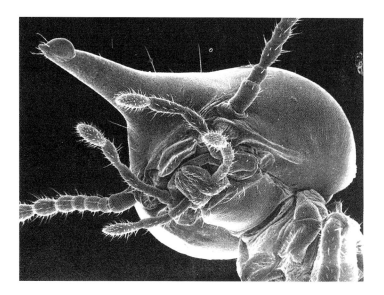

cause. His artificial ant had been glued down!

'The soldiers have conspicuous pointed nozzles on their heads from which they eject a sticky repellant fluid, produced by a gland that fills much of the cranial cavity. The secretion also acts as an alarm substance and draws other termites to the site. The target becomes surrounded by other soldiers, all spraying sticky fluid.

'The spray itself has a double effect on the enemy. Firstly it causes an irritation which results in the ant trying to clean itself. But the stuff has the consistency of rubber cement and it ends up simply caking itself with the glue. It then finds that every piece of detritus or sand in the environment sticks to it, and pretty soon the ant looks like a veal cutlet coated with breadcrumbs. It's just totally incapacitated.'

Several species of insect use this defensive technique, including those that cannot produce their own glue; instead they take secretions from plants, a form of behaviour which resembles tool use in higher animals.

The assassin bug

The most spectacular insect that uses this technique is a creature Eisner studied in the southwestern United States known as the assassin bug. This vividly coloured little beetle has developed an ingenious way of protecting its eggs from predators. The female scrapes resin from a particular plant and coats her belly with it. As she lays her eggs they too become coated with the resin. When Eisner analysed the resin he discovered that it contained camphor, one of the oldest insect repellants known to man!

Aphids and ants

Similarly, detailed nature-watching led Eisner to a series of extraordinary discoveries about a creature which human gardeners regard as their arch enemy: the woolly aphid. Firstly he revealed that the success story of this aphid was partly the result of a symbiotic relationship with ants. Ants guard the aphids from predators in return for 'protection money' in the form of the aphids' sweet honeydew secretion.

However one particular predator, the lacewing larva, seemed immune to the ant guards; Tom observed these little killers feasting on the woolly aphids right under the noses of the guardian ants. 'They were literally operating under the old ruse of the wolf in sheep's clothing,' he told us with delight. 'First they pluck the white wool from the body of the aphids and attach it to their backs as a disguise. They have even developed special hooks to hold the wool in place. Only when the larva is completely camouflaged and safe from the guardian ants will it go about its business of feeding on the aphids.'

Beetle weapons

Fascinating and ingenious though the lacewing larvae are, Tom's true favourites are the beetles. Given half a chance he will quote the comment made by the eminent English twentieth century biologist J. B. S. Haldane who, when asked what he considered had motivated God to create the world, replied: 'An inordinate fondness for beetles.' At the same time Tom believes that a more likely explanation for the success of these insects is their extraordinary mechanisms of defence.

'Beetles are, to all intents and purposes, chemical factories. They can eject the most unbelievable compounds from their glands to repel ants, spiders and centipedes; in fact all their major enemies. You just

continued

Some termites (*Cubitermes* species) carry their chemical warfare a stage further. They apply modified fatty acids, which are actually poisonous, to a damaged victim. Yet another group (*Rhinotermitidae*) have evolved a special mechanism for applying the toxic balm. Their labrum has been enlarged to form a bristle-brush which soldiers press against invaders.

A great number of insects have evolved chemical defence mechanisms of a complex nature, but only a few have employed them as successfully as the termites or placed such a reliance upon them. In some soldier termites their tanks of toxins exceed 35 per cent of their dry weight, causing one researcher to comment: 'For more than a hundred million years the insect world has had a chemical arms race.'

TOM EISNER
Insects' defences

Below: *Refined weaponry being demonstrated by a bombardier beetle. A toxic spray created from a mixture of chemicals explodes out of the beetle at boiling temperature from a special nozzle, to deter attackers. Eisner also demonstrated that the beetle squirts the spray in pulses, to allow microseconds of cooling for its own protection.*

wouldn't believe the compounds they make: highly concentrated acids, alkaline substances, and a range of poisons too.'

An hour's drive from Tom's laboratory in upper New York State is a beautiful canyon where we collected a bombardier beetle (which may also be found in the UK). Back in the laboratory, Tom attached one of these creatures with a blob of wax to a fine wire which he suspended over a sheet of chemical indicator paper. One gentle touch to the rear of the beetle resulted in a splash of colour fanning across the acid-sensitive paper. 'You see how it lives up to its name,' Tom announced with great excitement. 'It literally bombards! It's squirting a fluid from its rear that, believe it or not, is at the temperature of boiling water. It generates this jet of fluid in what is literally a chemical explosion.'

The nozzle used to direct the spray is also cunningly designed; in fact the bombardier has shortened wings so that the end of the abdomen where this revolvable gun turret is located can turn in any direction, and can even fire forward.

Having proved that the boiling spray is produced by a violent chemical reaction of two different substances inside the beetle's body, Eisner next pondered how the creature could contain the scalding liquid without cooking its insides. He suspected that this seemingly continuous spray is squirted intermittently, allowing gaps for cooling down. A scientist friend created a camera-flash mechanism that would allow the beetle to be filmed at several thousand frames per second. The processed film proved Tom right. At speeds many times faster than the human eye could detect, the bombardier beetle was indeed ejecting the spray in pulses.

One of the delights of being with nature-watchers like Tom Eisner is that you get caught up in the excitement of their obsessive curiosity. Nothing goes unquestioned. Take the case of the 'kamikaze effect' in ants that have been sprayed by a bombardier beetle. (Tom is very deliberate in his use of the word 'defensive' when describing insects' weapons; indeed it is most intriguing in a jungle as savage as the insect world that the majority of these mechanisms are defensive rather than aggressive.)

The bombardier beetle's spray may be red hot and highly acidic, but it is used in short bursts designed to deter. Curious to the last, Eisner noted that there were some ants which would never let go: the beetle would fire repeatedly until eventually the ant would be killed.

'We call this the "kamikaze effect" in the context of social insects such as ants. These individual ants can be envisaged as committing altruistic suicide. They sacrifice themselves, but as a result the beetle has depleted a substantial fraction of its reserves. We calculated that it takes five suicidal ants to deplete the beetle's supply. The beetle can then fall victim to, and provide food for, the remainder of the ant colony.'

Despite extensive research, Eisner emphasized that we know virtually nothing about the insect world. 'We have only just begun to scratch the surface. Until 50 years ago, most information about insects was of an observational nature. We knew nothing about how they functioned internally. We are now starting to apply a lot of experimental work to our enquiries

Bombardier Beetles:
A lesson in chemical warfare
Inside the body of the bombardier beetle a natural chemical warfare plant produces the highly repellent quinone, p-benzoquinone.

The chemicals are actually mixed on demand: they are stored in glands, then synthesized at the moment of ejection by oxidation of hydroquinones. The oxidizing agent is hydrogen peroxide.

The glands operate almost exactly like a pulse-jet engine. Each gland has two compartments; the inner and larger holds hydroquinones and the hydrogen peroxide; the smaller outer compartment – the reaction chamber – contains a mixture of catalases and peroxidases.

When muscle contraction of the inner chamber squirts the chemicals into the outer, there is literally an explosion. Gaseous oxygen provides a propellant which 'pops' the charge out at the enemy (with an audible noise) at a temperature of 100°C!

STRUAN SUTHERLAND
Snakes and spiders

Above: *Struan Sutherland: keeping an eye on the masters of overkill. One bite from a snake in this pit at a park near Sydney, Australia, could 'fill a squash court with dead guineapigs.'*

Below: *Sydney, Australia.*

about them, and if you couple that with the fact that only relatively few species have been observed in detail, and that there are literally hundreds of thousands of them, the insect world is still a great unknown.'

In addition, Tom pointed out that the insect world is suffering terrible attrition, largely because the slaughter and destruction goes on unnoticed and because humans regard insects as alien.

'We have to decide what our priorities are,' he stressed. 'Mine are very simple. I feel that there is so much still unknown about what is going on down there that it is our obligation to preserve this wilderness. I want my children to be able to go out and see things for themselves, not just to read about things that have already become extinct.'

STRUAN SUTHERLAND:
The poisonous creatures of Australia

As we have just heard, insects produce chemical venoms for their own defence. Sometimes the potency of these toxins is quite bewildering because there seems no sensible reason for the animal to produce such a powerful deterrent – they can be 'overkills'.

Australia is renowned for its array of insects and other crawling creatures dangerous to man: eight species of spider (although they are not insects we have included spiders because they are so fascinating), venomous centipedes and scorpions, and even a tick capable of causing paralysis.

Dr Struan Sutherland (who we have already met at the Commonwealth Serum Laboratory, Melbourne) is a world expert on these venoms. The creature of which he is most proud is the Sydney funnelweb spider (see page 105). The liquid that drips from its fangs is the venom for which there was no antidote until Struan Sutherland isolated its complex components in 1980.

It is said that the funnelweb is not only the most poisonous spider on Earth, but also the most aggressive, as our own contacts with this little creature confirmed. It does not, like most spiders, scurry into hiding as fast as it can go when danger approaches, but stands and rears into the attack position even when a movie camera that could crush it to a pulp appears. The fangs begin to drip venom almost immediately.

It is pure hard luck for Australians that their founding fathers chose the exclusive habitat of the world's

most deadly spider as the site for their largest city –
Sydney. About 500 people are bitten every year by
the funnelweb, due in no small part to the fact that
human habitats, like garages and outhouses, suit the
spiders very well.

Up until 1980 no one had been able to come up with
an antidote because the mix of toxins was so complex.
Then Struan Sutherland cracked this witches' brew.
The first odd thing about it is that it seems to affect
only humans and monkeys, neither of which may
sensibly be regarded as the natural prey of Sydney
funnelwebs.

Struan is also reasonably certain that the component
in the toxin which is so effective with men and monkeys is a bizarre 'extra'. 'It has so many other effective
components in its poison, it doesn't really need the
one that kills humans,' Struan revealed. 'There are
other odd features to the spider as well. The fangs are
huge with massive muscles driving them: they can
penetrate a child's fingernail, and you often have to
tear the creature off when it has embedded its fangs.'

Neither the mechanisms nor the ferocity are necessary for the everyday business of a funnelweb. Is it
even vaguely conceivable that somewhere in the past,

The World's Most Dangerous Spider
The Sydney Funnelweb (*Atrax robustus*)

Children have died within two hours of
being bitten by male funnelweb spiders.
The female is also dangerous, but not to
the same degree.

Other species of funnelwebs are
found in Eastern Australia, Victoria,
South Australia and Tasmania, but of
these only the Sydney and the northern
funnelweb are known to be dangerous.
The Sydney funnelweb is believed to
be limited to an area about 100 miles
from the centre of Sydney. It builds a
burrow (or will use a crevice) about a
foot deep. When mature, males leave
their burrows and roam, often into
houses, particularly after heavy rains.
They may live for eight years or longer.

Funnelwebs are one of Australia's
largest spiders. Males and females have
bodies 1–2 inches in length, but the
males are more delicately built, and
have two distinctive features: a spur
halfway along the second leg, and
finely-pointed feelers used to transfer
sperm to the female.

The venom (that of the male is five
times more toxic than the female's)
contains a low molecule weight toxin
called atraxotoxin, which attacks the
nerves of the body causing thousands
of electrical impulses to be fired down
them. The muscles of the victim twitch
convulsively, and there is a profuse
flow of perspiration, tears and saliva.
The venom also causes changes to
blood vessels, which can lead to shock
and coma due to brain damage.

Left: *Australians unhappily chose to
build their largest city, Sydney, at the
home of the world's most venomous
spider, the Sydney funnelweb. Until
Struan Sutherland cracked the witches'
brew of its venom, most bites proved
lethal.*

STRUAN SUTHERLAND
Snakes and spiders

on some strange rung of the evolutionary ladder,
these spiders found themselves in real conflict with
something having a similar metabolism and as large
as a human? That suggestion would be ludicrous
were it not for the specific man/monkey element in
funnelweb toxin. Most spider poisons cause little
more than discomfort in an animal as large as a human.
But you can be dead within 90 minutes from the bite
of a Sydney funnelweb!

Struan Sutherland may have identified the toxins,
and his laboratory now manufactures an antidote,
but the evolutionary motivation for so deadly a killing
ability remains a great Australian riddle. As we shall
see later, it is just one of many.

Struan Sutherland has made a considerable speci-
ality of this bizarre realm of nature-watching, and
during our visit introduced us to his recently published handbook *Venomous Creatures of Australia* in
which he lists what can only be described as an Anti-
podean miss-list (see left). Unquestionably, the most
fearsome creatures on this list are the snakes; proof
that aspects of the Australian fauna are employing
more venom than they can possibly need in the wild.
Top of the list is the small-scaled or fierce snake.

The fierce snake
The fierce snake is not actually fierce at all; it is shy,
reclusive and quite rare – but who can blame the
Australian pioneers for so naming it when it has the
strongest venom known to science? Fierce snakes
produce enough poison in one bite to kill at least
20,000 creatures the size of a guinea pig. That is, of
course, more food than any fierce snake could consume in a lifetime.

Its terrible reputation may have increased its
numbers, and the number of its victims. No living
specimen of the fierce snake had been found in the
wild this century before 1976, and in the 100 years
before that only three had been captured. Now every
zoo wants 'the most poisonous snake in the world'
and as a result they are quite common in captivity.
'It could be that the main use of our serum against
their venom will be for snake-handlers,' Struan commented wryly.

The taipan

In most Australian zoos, the display case next to the fierce snake is occupied by one or two silky, honey-gold taipans, labelled with their claim to fame – 'The Longest Venomous Snake in the World'.

'Taipans are very beautiful,' Struan enthused as he began to tell us the story of the snake-serum martyr, Kevin Budden.

The taipan is notoriously shy. It can sense the approach of a human from a great distance and will always hide. Scientists had great difficulty collecting a specimen until the brave Kevin Budden went into the Queensland sugarfields determined to 'bring one back alive' for science. The snake duly arrived at the laboratories, but not poor Kevin – he had been taipanned!

'As he was dying,' Struan recounted with great relish, 'he insisted that the snake should not be killed but sent down to Melbourne. There, two days later it was milked and good taipan venom was obtained for the first time.'

Too-powerful toxins

What can be the reason for these extraordinarily high toxin levels in Australian snakes? Most scientists believe that they could be the result of a reduction in the number of prey animals, such as mice and rats, as Australia became more arid and desert conditions spread. The snakes cannot afford to miss, so they have

Snake Senses

Snakes have re-designed some of the traditional five senses of smell, touch, hearing, taste and vision to a configuration better suited to a creature that crawls.

All snakes are deaf, but they can detect movement by sensing ground vibrations through their bodies. Actual vision is poor: some snakes are thought only to be able to see prey when it moves, but a great number of snakes also have infra-red sensing mechanisms – heat detectors – which in effect allow them to see in the dark.

The transfixing 'stare' of a snake results from the lack of an eyelid: snakes cannot blink. The eye is covered by a scale which is completely transparent other than when the snake is shedding its skin.

A snake's sense of smell (through its nostrils) is not very good but they more than compensate by taste/smelling particles picked up by the tongue (hence the constantly flickering tongue) which are analysed in two pits (Jacobson's organs) in the roof of the snake's mouth.

Snakes are all cold blooded and almost all of them hibernate in winter. In Australia, however, hibernation is rarely complete and a snake found comatose can 'come alive' and bite, very quickly.

All snakes can swim but, contrary to popular belief, none move that rapidly; a maximum of about 5 mph is the fastest recorded.

Left: *Elgant and deadly: taipans are the longest truly venomous snakes in the world and third in Australia's lethal league (see p. 109). The snake has especially long fangs and prodigious poison production – up to 400 mg per bite.*

STRUAN SUTHERLAND
Snakes and spiders

Venom

Venom is delivered by a highly efficient hypodermic injection. Poisons are produced in glands in the head. Delicate muscles squeeze controlled doses of poison into a channel in the fangs, to an exit point just short of the tip. Some snakes can rotate their fangs to allow a deeper, right-angled bite, and fangs are replaced when broken or worn out. It is thus possible for a snake-bite victim to display a confusing, single puncture wound. Some snakes bite, hold and chew to increase venom delivery.

Australian researchers have shown that venoms are complicated mixtures of poisons. The main component is, in most cases, a nerve poison (neurotoxin) but there are often additional toxins which can inhibit blood clotting, promote rapid dispersal of the poison and damage red blood cells.

Snakes are able to dislocate their jaws when swallowing, thus allowing the engorgement of larger creatures than would seem possible. Large snakes such as pythons can swallow small antelope and have sometimes been found with horns protruding from their (snake's) belly. The brain of a snake is completely encased in bone to avoid damage from struggling prey during swallowing. The food is moved down into the stomach by a series of teeth behind the fangs which curve backwards. Prey is almost always swallowed head first so that the hair and feathers lie flat.

developed over-kill. Interestingly it has been discovered that although snakes may be great producers of venom (and each species has a resistance to their own family poison), they have no more resistance than the rest of us to alien toxins. If an Indian cobra were to be bitten by a taipan, or vice versa, it would be in trouble.

Humans and snakes

Struan Sutherland has also made a detailed study of the case reports of human snake bites. They reveal, as he had always suspected, that the so-called victims (there are some 3000 a year, of which 500 require serum) could have avoided the incident if they had known more about the behaviour of snakes.

'Most bites are quite avoidable,' Struan insisted. 'People see a snake in the bush. It's probably asleep but they still attack it with a stick. The snake wakes up really angry, and frightened too. So they try to hit it again. This time the snake attacks and bites – of course. What else is it supposed to do?

'We've been able to cut down a lot on the number of incidents by giving expedition groups, particularly school groups, a little pep talk on how to avoid snakes.

'In recent years the fatality rates have dropped quite dramatically; simply, I think, because the management of snake bites in Australia is now the best in the world.' Researchers like the unfortunate Mr Budden and Dr Sutherland have altered the public attitude to the plethora of venomous creatures in Australia; that in itself is of great advantage to the snakes' conservation and they are now protected in almost every state.

Enigma variations

While we were in Australia, we had the opportunity to visit another nature-watcher who has also puzzled over the problem of an excess of venom in the tiny creature he has been studying for a number of years. This animal is featured, if a little out of context, on Struan Sutherland's list of venomous creatures. This list deals mainly with creatures we expect to be potentially dangerous; snakes, spiders and scorpions. But how about a cuddly, furry mammal that lays eggs?

These are creatures that nature-watchers have always found odd and out of place; you will probably guess that we are referring to the enigmatic monotremes.

The Lethal League

1. **Small-scaled or Fierce Snake** *(Oxyuranus microlepidotus)* Extremely rare, found in small patches over a large area of southwestern Queensland, western New South Wales, the Northern Territory and South Australia.

Average venom yield 44 mg, maximum ever recorded 110 mg. (More than 100 times as toxic as the North American diamondback rattlesnake, enough venom produced in one load to kill 250,000 mice.)
Identification: Glossy black head, deep brown above with cream belly. 3–6 ft long.

2. **Common or Eastern Brown Snake** *(Pseudonaja testilis)* Common and highly venomous; average venom load 67 mg (one dose kills well over 100,000 mice). Found in the whole of eastern Australia, and (except for the mainland tiger snake) account for most serious snake-bite attacks on humans. Venom extremely neurotoxic and rapidly upsets blood-clotting mechanisms. Unlike many snakes it is highly active during the day in hot conditions, and after biting often holds on to ensure maximum venom delivery.
Identification: Chocolate-brown to blackish, shading to orange, 5–8 ft long.

3. **Taipan** *(Oxyuranus scutellatus)* The longest venomous snake in the world (and many think the most beautiful), with the longest fangs of any Australian snake (13 mm) and equally copious poison production (maximum 400 mg) with high toxicity.

The taipan is an inhabitant of the northern coast of Australia, down to Brisbane. (Thought to occupy a wider territory, but the snake found in the west has since been identified as the fierce snake.)
Identification: Light or dark brown above, creamy-yellow belly, reddish or pink spots towards front and 'mother of pearl' bloom to belly skin; average 7 ft long, longest found 11 ft.

4. **Mainland Tiger Snake** *(Notechis sculatus)* Common in densely populated areas of southwest Australia: as a result is most common cause of snake bite in Australia. They hunt frogs, rats and mice on warm summer evenings, along banks of dams and in and around houses. Several fatal attacks have occurred on lawns after dark. Have fourth most poisonous venom in the world, with average yield sufficient to kill 40,000 adult mice (with one bite).

Venom made up of a number of separate toxins, including three neurotoxins which inhibit breathing and promote muscle damage resulting in kidney failure and death.
Identification: Broad head, colour varies from pale brown to near-black, and while the classic tiger snake has 45 yellowish bands along its 3 ft body, some have none.

5. **Death Adder** *(Acanthophis antarticus)* Two species exist; *A. pyrrhus* confined to the arid area of central Australia and thus rarely encountered by humans; and *A. Antarticus* (Common Death Adder) widely distributed throughout most of Australia and known to produce huge quantities of venom (235 mg) promoting severe paralysis. Fifty per cent of victims die before an anti-venom becomes available.

Believed it strikes only when touched. Then attack is very swift and normally successful as snake has very long fangs (average 6.2 mm) and ability to align them at right angles to ensure deep bite.
Identification: Resembles a viper with a broad triangular head and short (3 ft) body. Colour varies from light brown to near-black with, generally, bands of lighter colour across body; yellow or grey belly.

7. MONOTREME-WATCH

There are just three species of monotreme on earth: the Duck-billed Platypus (*Ornithorhynchus anatinus*) and the Common Echidna (also called Spiny Anteater, *Tachyglossus aculeatus*) in Australia, and not far away, the Long-beaked Echidna (*Zaglossus bruijini*) of Papua New Guinea. The family is called 'monotreme', which means 'one hole', because they have only one opening in the lower body through which their eggs are laid and body wastes pass. At first glance, and in this more knowledgeable age, such an identifying feature would hardly seem to justify the turmoil in biological science which took place on its discovery almost two centuries ago. On the other hand, in all that time, we have still not come across another mammal which lays eggs.

To confuse the issue further, monotremes are also marsupial: they have a pouch. Once their egg has hatched they revert to what we regard as more typical mammalian behaviour by suckling their young in the pouch. The milk is produced by mammary glands, but there the similarity to other mammals stops. They have no teats; the milk emerges directly from glands at the base of a little patch of hair.

It would have been impossible for us to go to Australia without taking note of the present status of these unusual animals. At the University of New South Wales we asked Dr Mike Augee, one of the few people who knows about monotremes, to tell us more about them.

Opposite: *Mike Augee has a special interest in the most enigmatic animals of Australasia, the monotremes: the duck-billed platypus and, opposite, the common echidna. Julian Pettifer and* Nature Watch *saw both with him on an expedition to Kangaroo Island.*
Below: *Kangaroo Island, Australia.*

MIKE AUGEE:
The shy monotremes

The discovery of the duck-billed platypus

The duck-billed platypus
There are a number of animals in the world that would appear to have been assembled by a committee. The African wildebeeste has often been so described. When the particular 'committee' sat down to plan the platypus it is almost as if it started work on a bird, lost its way, mislaid some of the parts (they have no teeth), and decided in the end on a mammal. When it

MIKE AUGEE
Australian monotremes

added a beak to the otherwise furry platypus the committee had produced a creature that was bound to provoke the scientific uproar which greeted the arrival of the first specimen in Britain.

'The first one was sent in about 1798,' Mike Augee explained, 'and at the time it was believed to be a scientific hoax cooked up by colonialists. This re-action was forgivable because at that time there were a lot of these "wonders" being shown at country fairs in Europe. The most famous of these was a dried "mermaid" which was made by stitching the head of a small Indian monkey onto the body of a fish.

'This, and other specimens arriving in England at that time, consisted only of skins; no organs or skeleton were present. Naturalists were bewildered when they were presented with a creature that appeared to have a duck's bill, but was covered with some sort of leathery stuff; a tail that looked like a beaver's, and claws that looked like a mammal's yet were webbed like a duck.

'Finally, an expert at the British Museum, George Shaw, was called in to examine the creature. He had made his reputation by exposing the Indian mermaid

Below: *Duck-billed platypuses are furry, aquatic and, of course, duck-billed: echidnas are spiky, land-bound and have a tube through which they suck up ants. Their skeletons, however, show an almost identical animal re-shaped to the de-mands of its chosen habitat. (Platypus in foreground, echidna at rear.)*

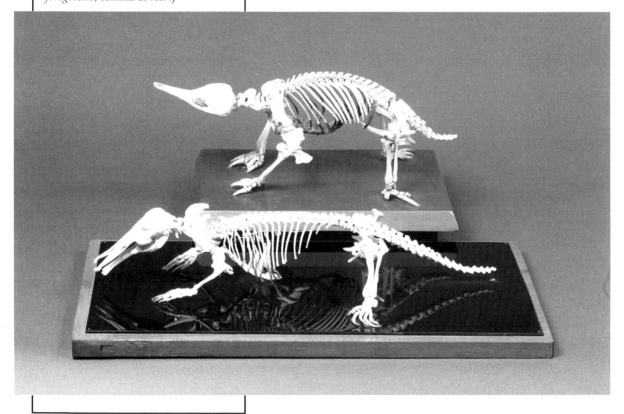

as a hoax (he found the stitching round the neck). So the first thing he did with the specimen was to search for signs of artifact. When he was unable to find anything suspicious he reluctantly announced that it might be a new animal.'

The arrival of whole specimens helped the investigations considerably. But then people began to realize that their reproductive behaviour was even odder than their appearance.

'Sir Everard Home, the first man to dissect a platypus, said it looked like it might lay eggs, but nobody really believed him, and it was another 50 years before the dust settled,' Mike explained. 'Finally, in 1884, a Cambridge zoologist called W. H. Caldwell was sent to Australia specifically to settle the question. He found a platypus with an actual egg tucked away in its pouch, and was so excited he telegraphed his findings to a meeting of the British Association in Canada: "Monotremes oviparous, ovum meroblastic" – there is a mammal on Earth that lays eggs!'

Australia is rightly proud of these special creatures. Pioneer conservationists organized a total ban on the hunting of the platypus in the 1920s. Today it is very well established in most of the larger river systems of the east coast of Australia, particularly in the Snowy Mountains and inland New South Wales.

'Both the platypus and the echidna lack teeth; the most definitive characteristic which links them inexorably and uniquely together is that they are both egg-laying mammals.'

'You can only really appreciate how closely related they are when you study their skeletons,' he agreed. (See page 112.) 'You can see how the so-called duck-bill extending down from the skull is made up of the same bones that form the cylindrical snout of the echidna.'

The monotreme expedition

When we first suggested to Mike Augee that we would like to see these animals in the wild, he tried to talk us out of the idea. He told us that the reason very little is known about them is because they are difficult to even catch sight of. A ripple on the water is all that a platypus may offer, and although echidnas are easier to watch as they are land animals they are very shy. (Mike Augee only achieved his seven-year field study of echidnas by radio-tracking.)

MIKE AUGEE
Australian monotremes

MIKE AUGEE
Australian monotremes

At last, persuaded by our determination, we set off with Mike for the Flinders Chase Reserve on Kangaroo Island (south of Adelaide), which has been offering sanctuary to a range of Australian fauna since 1919. It is popular with tourists, who come to see the two species of Australian monotreme, and also a race of extraordinarily tame western grey kangaroos.

Finding an echidna

It took some time to locate the first bleep, indicating the presence of an echidna. Mike's radio equipment is limited by the fact that he insists on using the smallest possible transmitter pack. 'The mistake a lot of people make with this method is that they end up with more technology than animal. We're using the smallest transmitter we can, to interfere as little as possible with its movements.'

We looked behind bushes, in groves of blue gum trees, and in dead logs. Finally, with Mike not a little embarrassed, we stopped before a bleeping molehill. 'Do you want to dig him up?' he invited. 'I have to change his battery anyway.'

We let him do the digging the first time. The second spadeful brought up a sandy tuft which shook itself out into what looked like a cigar-smoking giant hedgehog. This creature gave us a menacing glance, walked round us in a tight circle, then buried itself back in the hole we had excavated.

Echidnas have a peculiar walk. They roll along, much as you might imagine the slow, serious Eeyore does in his wanderings with Winnie the Pooh and Christopher Robin on their way to have tea with Owl. Our photograph can only hint at this steady, studied perambulation.

'There isn't a lot to see,' Mike confessed. 'I did warn you. They don't have much to worry about so they don't do much. There is no animal in the living fauna that can attack an echidna when it's buried. So they walk about in their funny way, pick up their food, and at the first sign of anything unusual (like new batteries) they dig in at an incredible speed. We could dig him up again if you like, they don't seem to mind, but he'll go straight back down again.'

An echidna's life

We sat on a log, hoping the echidna might dig itself up, while Mike told us what he knew about their habits.

'They move about within a definite area. I'm very

careful not to say territory because that, to most people, means that an animal has a specific little plot it defends. That certainly isn't the case with the echidna. They just seem to use a plot of land which must somehow be definable to an echidna. We have labelled it their "home range". These usually overlap.

'There's no evidence of any sort of hierarchy between echidnas, or any sort of defence mechanism,' Mike said. 'Echidnas are very tolerant of other echidnas, to the extent that we can never be really sure that we have captured and put transmitters on all the animals using an area. In this area of study there are as many as seven individuals with overlapping home ranges.'

Above: *Echidna behaviour has so far proved singularly unexciting. Those we met appeared to have one characteristic in common, an insatiable curiosity.*

Field studies are also complicated by the fact that the animals live a very long time. Echidnas in zoos have survived for 40 years. Mike Augee's seven-year study can only provide what he feels to be an 'educated guess' about how long they survive in the wild.

MIKE AUGEE
Australian monotremes

MONOTREMES – the Mystery
Mammals

There are some 4070 species of mammals, the broad band of creatures which took over from the dinosaurs after an inauspicious beginning in the Triassic period some 200 million years ago. By the late Cretaceous period, 75 million years later, they were the most successful vertebrates.

Most experts believe that the secret of the success of mammals is lactation – their ability to care for their young with a ready food. Young dinosaurs (like modern day crocodiles) were born as tiny replicas of their parents, entirely dependent on the food nature provided.

None of this sheds much light on the evolution of the monotremes; these mysterious egg-laying mammals with affinities to birds, who are the last surviving examples of a mammal sub-class, the *Prototheria*.

On the ladder of mammalian evolution there are three distinct steps; *Prototheria* (egg-layers like the platypus and echidna), *Metatheria* (marsupials or pouched animals like the kangaroo), and finally *Eutheria* (placentals – from mice to humans).

There is a 200 million-year evolution separation between the egg-laying *Protheria* and the live-bearing *Theria* and, at the present time, an almost total dearth of information which might cast any light on how mammals progressed to what is seemingly a more successful method of birth and infant care.

'They only produce one young a year, which is typical of a species with a relatively long life span. Another indicator of longevity is their very low metabolic rate. They are exposed to a number of parasites in the wild, including ticks, and a flea that's specific to echidnas; they probably live about ten years.'

In an attempt to learn more about echidna behaviour, Mike set up a captive close-study environment as near to an echidna paradise as he could. 'It was a large complex enclosure. We moved their food around on a statistically-arranged programme to try and get them to establish some sort of territory. They just wouldn't do it! We had dirt and nice hollow logs but when the sun came out you'd find the whole lot of them in a pile on a piece of concrete which, to us humans, looked totally wrong for an echidna. So the only thing we can say about them after all that study is that they're unpredictable and very, very tolerant.

'I can tell you what they eat. This is quite interesting because they have a long sticky tongue that is completely specialized for eating ants. Echidnas are quite choosy. They particularly like meat ants, but only in the spring. If you pound the ground the worker ants pour out of their nests in great numbers, but at all times of year except spring the echidnas ignore them because the workers have a really nasty sting. In spring, however, the echidnas just tear the nests apart, and in spite of the fact that they become covered in worker ants they dig down into the lower galleries to get at the female ants, which are then full of fat. There's also a nasty thing around here called the bull ant with very large pincers, which they avoid.'

Our buried echidna was showing no signs of re-emerging. 'He might well be feeding down there,' Mike said. 'They don't mind being buried. Obviously they don't mind dirt and they take in much more soil than ants when they're feeding. In fact one of the problems with keeping echidnas in captivity is the need to feed them a lot of bulk to replace the dirt they would be eating in the wild.'

By now the sun had dissolved into a spectacular Australian sunset and Mike suggested we visit a river and some ponds where he knew platypuses were living. We set up our camera on the bank and watched the light fade with southern hemisphere abruptness.

A ripple and . . . a duck-billed platypus

Suddenly Dr Augee was on his feet, pointing at a spot in the river. We stared hard but saw only a ripple, as if a stone had been thrown close to the bank on the other side of the pool. (Fortunately, however, modern film cameras fitted with high-speed lenses can outdo the human eye, and we had captured a rare metre of film. When the film was processed it revealed a creature, looking not unlike one of Charlie Chaplin's famous boots, being propelled across the surface of the pond by the leisurely flapping of its beaver-like tail.)

'That may be your lot,' Mike said. 'A good sighting

The Rearing of the Young Platypus

The platypus builds very elaborate burrows, 20–60 ft long. Tortuous passages are excavated through the river bank to the nesting chamber, which is then protected by a number of blocks built by the female.

The nest is rounded and built of grass and leaves: one, two or three eggs are laid, the usual number being two. The eggs measure about 14 mm, are soft-shelled and off-white in colour. When a single egg is laid, it is usually a little larger than the twin egg.

When hatched, the young are suckled by milk secreted from a number of small pores (not teats) in the abdominal skin.

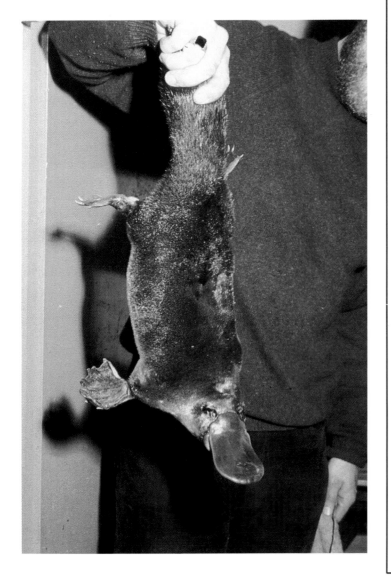

Left: *Fauna or Fake? That was the question troubling science after the first platypus specimen reached Britain in about 1798. Mythical creatures, like mermaids, were faked for freak shows by sewing small monkey heads on large fish bodies and the first test conducted on the platypus was an attempt to pull off the 'fake' duck beak. When their reproductive behaviour (egg laying) turned out to be even more bizarre than their appearance, a special expedition was mounted to clear up the mysteries of the monotremes.*

MIKE AUGEE
Australian monotremes

Above: *Unresolved mystery: no viable purpose has ever been found for this bizarre platypus feature – a highly venomous spur on the ankle of each hind limb.*

in platypus terms. A friend of mine has been trying to track them down for years using transmitters, as we've been doing with echidnas. But the only place you can attach it to is the tail and the signal fades as soon as they go into their burrows.'

A platypus' life
So just how much is known of the behaviour of what must surely be regarded as the world's most extra-ordinary land animal?

'Very little,' said Mike reluctantly. 'I can't even tell you whether whole families live together in the burrows; we suspect that males and females occupy solitary burrows, but possibly the females will share with their young until they become independent. The burrows are quite complex and end in chambers where there are nests made of twigs and bits of grass; it is quite common to find two eggs here.' (When hatched, the young feed on milk from the mother's mammary glands – there is no pouch – for 3–4 months.)

Several Australian zoos have now built the expensive facilities required to keep platypuses in captivity (none abroad have been successful), and much has been learned about their feeding behaviour. The large flat bill is used as a dredging tool in the river bed. When they submerge a third, waterproof eyelid comes down and the platypus blindly hunts for its diet of crustaceans and earthworms. Early researchers thought the animal must be using some form of sonar detection, like dolphins do, but it is now known that the bill is covered with shiny black skin containing many sensory nerve endings which can detect the movement of prey.

Like their land-based cousins, the echidnas, platypuses take in a lot of dirt and silt when they dredge for their food.

'It's like panning for gold,' Mike illustrated. 'They have pouches in their cheeks to separate out the organic bits; they chew and swallow these and spit out the sand and dirt.'

The sting
One of the most intriguing features of the platypus is its defence mechanism; the male has a spur, a clever hypodermic that delivers a low-level, but very toxic, poison.

'The spur is on the ankle of each hind limb. It's not a toe but a separate structure entirely, connected by a duct to a poison gland beneath the knee. It can deliver a really hefty dose of poison. In laboratory experiments done some years ago, the venom from a platypus gland killed a rabbit. Humans have occasionally been spurred: the victim suffers stiff joints and considerable pain.'

After more than a century of close study, no one has come up with a satisfactory explanation for why this small creature, almost wholly unaggressive and free from any serious predation, should maintain such a venomous weapon. If the terrible overkill of some Australian snakes and spiders seems strange, the inexplicable venomous spur of the male platypus is even stranger.

There are suggestions that it plays some part in courtship. Mike Augee disagrees and cannot believe that the males have really developed something so nasty merely in order to subdue females. The theory arose because during the breeding season, under the influence of the male hormones, the gland enlarges. Mike postulates that the spur is a weapon of threat used when males establish territory in the breeding season, and that this may be a method of spacing them along the banks. But this cannot explain the strength of the venom. In all the years of study, no one has ever seen this dangerous little weapon in use, even though 'rather violent tussles between males' are not unusual and have been closely observed.

(In echidnas, there are structures similar to these venomous spurs, except that here they are not functional – the echidna's venom duct and gland are degenerate.)

8. LION-WATCH

We are often asked by people to recommend where to go to see the best animals. It is the most difficult question anyone could ask because the world has such a fantastic variety of things to offer. If you can find sufficient money to fly to Africa we would urge you to follow the example of vast herds of East African animals and go to the Maasai Mara National Park in Kenya, a little over an hour's flight from the capital, Nairobi. In 650 square miles there are 450 different species of bird and 95 species of land animal. There are many different kinds of habitats, a huge array of herbivores and all the attendant predators – cheetah, hunting dogs, lions, jackals and hyenas.

JONATHAN SCOTT:
Wildlife in the Maasai Mara

Jonathan Scott heard about the Maasai Mara from the television. 'I used to sit in front of it transfixed by the wonders of Africa. Then I began to read about Africa as well, books like *Jock of the Bushveld* (Sir Percy Fitzpatrick, 1907), which rekindled memories from another era.

'Suddenly the idea dawned on me that it would be fascinating to go out there and do that myself. Of course I never dreamed I'd make it and I can't really believe that I'm sitting here today.'

In 1974 he managed to save up enough money to join an overland party from England to South Africa which passed through the Mara. 'On seeing it I knew that this was the place I had to come back to – this was where I could do what I really wanted to do, which was basically to watch animals.

'I recall the time I came to Mara knowing that I might actually be able to stay and work here. I took a drive out to the river and immediately saw two magnificent male lions. That was enough, I knew I was in Paradise!'

Why Scott became a wildlife artist
Jonathan Scott knew he had also to earn his daily bread, and he revived a long-neglected skill with pen and pencil, drawing the animals that surrounded him

Above: *Jonathan Scott: drawing animals to understand them better.* Below: *Maasai Mara, Kenya.*

Opposite: *Dawn yawn – Maasai Mara lions greet another day.*

JONATHAN SCOTT
Maasai wildlife

The Maasai Mara Game Reserve is a 650 square mile area of high plains grassland which is part of the Serengeti ecosystem. It is owned by the Maasai themselves and they run it jointly with the Kenyan government. Along with the Serengeti, it is the last place on earth where you can still see such an abundance of plains game – literally millions of animals on the move including, for obvious reasons, a very heavy concentration of predators.

in such profusion. A Nairobi businessman got to hear of this reclusive young Englishman sitting in the Mara, painstakingly assembling impressions of the animals via the pointillism technique of thousands of fine dots. His admiration for these illustrations resulted in the private printing of a limited edition of East African birds of prey. When we met him, in September 1981, Jonathan was completing the illustrations for his first book *The Marsh Lions*, with text by another Mara enthusiast, Brian Jackman.

We think you will agree that Jonathan's drawings go far beyond photographic images. This results from his essential attitude to pointillism drawing as an extension of photography, at which he is also very expert (as you can see from the photographs). 'Anyone can take a photograph of a buffalo and end up with an image of a buffalo, yet understand very little about the animal. I have found, from drawing a buffalo's ears or horns, that you can't really know until you try to get them right on paper. I hope ultimately that I can convey to people the sense of what a buffalo is in terms of its very essence – the massive animal, the fearsome creature.

'I really love it. There are long periods when I'm on safari and don't draw. Then when I sit down again I realize just how much it means to me. It increases my understanding of animals and allows me to become absorbed completely in them.'

Perhaps because of this, Jonathan has become one of the most perceptive nature-watchers we have met. His insights appear to go beyond the observations of field biologists who pride themselves on their studies of behavioural detail. For him nothing that moves in the Mara is without meaning. It is as if he has been trapped in the web of life there and is sensitized to the deft interplay of natural events. This is best illustrated in practice by his ability to find animals, particularly predators, in the Mara. At first we thought he had simply developed extraordinarily good eyesight. Then we realized we were being led to a pride of lions or cheetahs, that he could not possibly have seen! Jonathan makes his way around the Mara by watching what everything else is doing, sign-reading if you like, but at the most sophisticated level, aware of every movement.

His favourite spot is a damp patch near his camp which he has called the 'Marsh', and although his most concentrated nature-watch has been of the lions which once lived nearby, we suspect that his true love may be the birds of this marsh, in particular the pied kingfisher.

When we first saw this bird it appeared to be fixed in the air, hanging on a string in a vibrating mirage. Slow-motion film showed that the mirage was an incredibly fast wing beat. The kingfisher could sustain this effect of defeating gravity for almost half a minute, delicately adjusting the position of its body to take account of the gusting wind. This huge expenditure of energy was for one purpose – watching fish. After hanging in the air, the kingfisher would plummet into the marsh like a stone, emerging on most occasions with a catfish.

'This is the only species of kingfisher to hover regularly,' Jonathan pointed out. 'Having caught the fish they take it to a special rock and whack it to death. Then they manipulate it round to a head-first position – it is always head-first – before swallowing it.'

Marching around the fringe of the marsh were several birds which could have walked straight out of a cartoon. Their general appearance was that of giant, scraggy crows with necks daubed blood-red. They had a heavy, rather lugubrious walk and what appeared to be a nervous twitch: the head was thrown sharply back every few minutes. At first we thought we were also near lions but the low drumming sound we could hear was coming from these birds.

'Ground hornbills,' Jonathan grinned. 'Ostriches make a noise like that too. The head flip is a feeding action. The hornbills walk round the outside of the marsh and through the grassland, picking up grasshoppers or dung beetles and other large insects. You'll see them pecking through the elephant droppings for tasty morsels. Then they flick their heads back, open their bills and the unlucky specimen drops into the back of the throat.' Every bird has something special going for it as far as Jonathan is concerned: the mighty fish eagles call 'with the cry of Africa'; hammerkop storks 'are thought to bring bad luck'; many northern birds which migrate to the marsh are 'old friends from the UK'.

A day out with Jonathan Scott

The following morning we rose at dawn and went in search of the inhabitants of his paradise; our first contact was perhaps the most surprising of all. Driving up a narrow ravine where Jonathan had once photographed a large leopard, we spotted a creature which looked like a giant guinea pig but

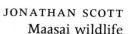

JONATHAN SCOTT
Maasai wildlife

Above: *Pied kingfisher reviewing the prospects of a meal. This patch of marsh, a mecca for dozens of different birds, has been the focus of Jonathan Scott's attention for a number of years.*

JONATHAN SCOTT
Maasai wildlife

turned out to be the closest living relative of the elephant! It was a rock hyrax, the cony of biblical times. Evidence for their extraordinary kinship with the largest living plains animal is to be found in their feet, although we were hard-put to see even that similarity. 'On their feet they have sort of hoof-like nails,' Jonathan assured us. 'These are indeed rather similar to elephants.'

There were no leopards in the gorge that day; in fact they eluded us throughout our stay. Even in the lush Mara which is protected round the clock and

Above: *Leopards appear to be making a comeback in the Mara after a decade of heavy poaching. A world-wide campaign by conservationists has severely restricted the trade in spotted cat skins. The chance to photograph one of these shy beasts, however – especially looking you calmly in the eye – comes rarely.*

patrolled seven days a week by people like Jonathan Scott, poaching has made leopards very scarce. A decade ago there was terrible attrition all over Kenya and neighbouring Tanzania but today the position has improved considerably. Unfortunately, this is partly because there are so few predatory leopards left. The killing of leopards was a particularly brutal activity; poachers went to bizarre lengths with red-hot spears to kill the animals without damaging the valuable skins. Those leopards that have survived have become highly secretive and, of course, are nocturnal hunters.

The cheetah

That other beautiful hunting cat, the cheetah, is seen by almost every tourist, but in Jonathan's opinion there are probably more leopards than cheetah. 'They are both solitary cats,' he explained, 'but the cheetah hunts by day and is thus seen more often.'

We stopped to consider what appeared to be a tranquil scene, a female cheetah with a large cub, but Jonathan watched the adult for a while and said: 'She's limping; that spells disaster for a solitary cat like a cheetah. Social cats like lions can rely on others of their pride to do the hunting and supply food, but with a cheetah you either kill your own prey or you don't eat. Cheetah rely almost entirely on their stealth and speed. They are the fastest animals on four feet. If they suffer an injury their prospects really aren't that good.'

We wondered in these circumstances whether this also bode ill for the cub.

'Not necessarily,' Jonathan replied. 'Another aspect of living a solitary life is that the young mature faster than say, lion cubs, who have the support of their pride and can afford to spend more time growing up and learning the skills of hunting; cheetah young don't have this luxury. If that cub is 18 months old it can hunt and will be able to support itself. A lion cub of that age will not be able to, even though it would have started to learn how to hunt.'

Maasai giraffe

We rolled on across the waving wheat-grass, tinted red in both dawn and evening light. We chose the sparkling waters of the Mara River, fringed with thick green trees, to have our lunch. These trees were most peculiar; they all seemed to have suffered a pudding-basin haircut! 'It's the giraffes,' Jonathan smiled. 'This is known here as the "browse line". It's dictated by the tallest of the giraffes. They come along and eat up a tree until they can't reach any more foliage, then they shove out their long tongues, 18 inches long in some cases, and give the line a final trim.'

The Mara is graced with its own unique subspecies of giraffe, the Maasai, and in a glade deep in the riverine forests, we saw a peculiar sight; three giraffes 'necking' so violently that they almost knocked each other off their long, thin legs.

'You don't see it happening very often. When it was

JONATHAN SCOTT
Maasai wildlife

Above: *Necking – a term of affection surely invented for giraffes – actually occurs between these tall ungulates and could be part of courtship rituals, or at least the establishment of dominance amongst the young males.*

first observed, naturalists assumed that it was some kind of amorous behaviour between males and females. We now believe it is play confined to males, especially when they are in the sub-adult to early adult stage, to sort out hierarchies. But they really go at it; banging their rumps together, swinging those long necks and huge heads against each other and giving themselves a general buffeting. It establishes who is the top dog and is very good for the stability of the group because, by establishing a dominant structure early on, you get very little destructive fighting'.

Hippopotamuses and . . . a lion
The only worrying moment on this expedition, so far, had occurred in the middle of the previous night when a hippo strode straight through our camp, taking our portable lavatory for a walk. The noise had been incredible. Perhaps we should have taken that incident as an omen and not got out of the car to visit the hippo pools in the river; not that the hippos caused the problem.

Most game parks do not allow you out of your bus or car, and if you are touring without a guide you should make it a rule never to leave your vehicle. Sanctuary animals have grown remarkably used to cars and you will see more if you stay inside, because most creatures panic if they suddenly see an alien human outline.

In the Mara, your guide lets you out for a stretch at one or two safe sites, including the high bank overlooking the hippo pools. Because we had professional status and Jonathan as our guide, we walked down the bank and took some very close-up shots of those 'aquatic monster pigs'. The hippos are a noisy lot, and we had just grown used to their grunts, coughs, sighs and 'raspberries' when suddenly Jonathan stood stock-still and indicated that we should do the same.

'There's a lion here,' he said flatly, like a medium who has just detected an alien presence. 'Just stay put while I locate it.' Cameraman, Noel Smart, located the lion with his 600mm telephoto lens. We heard the slight whir of the camera starting and Noel's matter-of-fact: 'The camera's running. It's in the grass on the other bank. There's a female there too. Nobody move, it's looking straight down the lens.'

We were not intending to move. Julian, making up our advance guard, asked half-jokingly: 'Do lions swim rivers, Jonathan?'

'They do.'

The golden rule of nature – when in doubt, retire – held good, however, and in a few minutes we were alone again with the big wet hippos. Later Jonathan told us more about the hippo, the animal responsible for the most deaths among the local people. 'They're not carnivorous, of course,' he explained. 'They come out of the water at night and browse the grass. But they don't like being disturbed, or crossed, and they pack a terrible bite. They can run a lot faster than we can, and that's three tons of hippo on the hoof.'

In the milk chocolate swirls of the Mara River, the huge hippos continued to bob up and down, opening the flaps that sealed their nostrils when under water

to blow air in our general direction. 'I always think of them as Cinderellas. In the first light of morning you find them hurrying back to the security of the water.'

Jonathan was with us when a huge pink hippo demonstrated that it is possible, in the Garden of

Above: *Wary hippos in the Mara River. Heavyweight Cinderellas (at night they leave the river to graze but invariably return before dawn); Jonathan Scott is particularly wary of these animals, pointing out that they account for more deaths among local Maasai than any other creature.*

JONATHAN SCOTT
Maasai wildlife

continued

LIONS

The world's present lion population is in fact a slim residue of the *Felidae* family, which once explored much of the world, as testified by cave paintings of European lions (15,000 years ago) and references to their existence in Greece (by Aristotle) 300 years before the birth of Christ.

The lion genus (*Panthera*) now includes five species. They are found in most parts of Africa and there is one small population in northwest India. The lions of the Maasai region have been defined as a separate subspecies, *Panthera leo maassaieus*.

Lions are massive cats, with a body length sometimes exceeding 10 ft (excluding the tail), weighing in excess of 500 lb. Gestation is short for such a large animal (something over 100 days) and as a result, lion cubs are tiny and need the close social protection of the pride.

Prides range over an area of 10–150 square miles. In the Mara, where game is plentiful, these territories are usually much smaller. A pride is in fact a loose-knit collection of groups (subprides), even a number of 'companionships', occupying the pride-range.

There are normally 4–12 related females forming the vital core of every tribe. Surplus females (prides establish an optimum size depending upon the density of food prey and other lions) are driven out when 2–3 years old and are normally forced to live a solitary life, until adopted by a pride later. Young sub-adult males usually leave of their own accord (or are driven out) at about the same age as the surplus females. They then live a bachelor existence for a year or two until they are able to take over a pride and become the breeding males. Their control of their new pride then depends on how well they can drive off rival contenders, but there are records of males ruling the same females in a pride for a decade.

Eden, for 'lions to lie down with lambs'. The 'lion' in this particular drama was played by a hippo – composed, sedate and all-powerful, but a little late getting back to the sanctuary of the river. Commanding her chosen entry point was a large crocodile which the hippo nosed contemptuously. Instead of the confrontation we were expecting, the crocodile moved back a few feet to allow the hippo room to lie down and take a nap.

In the cool of the late afternoon there were signs of stirring under the scrubby acacia trees and in the dusty dongas alongside the river. It was time for lions, and lion-watchers, to be thinking about the business of the evening.

The Marsh Lions

In 1979, soon after coming to the Mara for his first extended stay, Jonathan began to study a group of lions (his so-called Marsh Lions) which virtually set up camp round the large wet patch (the Musiara Marsh) in the middle of the park, during the period of the migration. Since then he has been making detailed notes and photographing and drawing this same group, which has given him a remarkable insight into the behaviour of lions. Pure nature-watching has grown into a genuine relationship that he now admits is compulsive.

'I have watched some of these lions since they were cubs. I've seen them develop through the whole pride system. Some have had to leave the pride to try and make a living for themselves.' Every one of these familiar creatures has, as he puts it, 'a story in its face'.

He has come to appreciate the role of the large males in the pride, and now questions the popular view that they play a minor role to the females who raise the cubs and do most of the hunting.

'The males fulfil a very important function – they ensure the integrity of the territory in which the pride is operating. When lions are in a pride situation and are defending a territory, the size of the males with their huge manes, together with their constant marking of the territorial boundaries, says to other lions who might otherwise try to break up this stability, "this ground is occupied, don't come near here." '

Within this framework of stability, the females can reproduce and raise the cubs in considerable security. There is complete chaos for a while if the males are lost from a group. Sometimes pride males may be driven off by other males, as happened to the Marsh Lions. Then they live solitary lives.

Male lions also contribute to the wellbeing of the cubs in certain situations. They will often allow cubs to feed with them while denying the females. In hard times this means the cubs get more food than the hungry females would allow them.

The wildebeeste migration

Just one event in the Mara overshadows Jonathan's obsessive interest in the lions: the annual wildebeeste migration. Every February, with seasonal regularity, about two million wildebeeste, a quarter of a million zebra, antelope, gazelle and their accompanying phalanx of predators and scavangers cross the Mara River to get to the lush grasslands of the north. 'They comprise the plains game of the mighty Serengeti,' said Jonathan with real awe. 'That's directly south of the Mara; in fact, the two are regarded as part of the same ecosystem. When this mighty population recognizes that conditions are getting too dry in the Serengeti, it heads north in search of greener pastures. Tens of thousands of animals come crashing into the waters of the Mara every hour. It's a densely packed mass of bodies. The horizon is literally black with animals. From the air, it appears as if a huge swarm of black ants is pouring down the ravines into the river like a flood of oil. Up to 10,000 of them never make it, in particular the wildebeeste. They throw themselves off the river banks into a teeming confusion of animals that are already in trouble in the fast-flowing water. Getting to the other side irrespective of the risk is their imperative. Those that make it across have an even greater problem on the other side because there are so many animals trying to get out. They get swept away, and countless numbers die when they get trapped against dead tree stumps and dams made up of the bodies of earlier victims. Then the vultures spiral in the sky, and drop down to gorge themselves on this bonanza of fresh meat. It attracts the hyaenas, jackals, crocodiles and, of course, the lions.'

Jonathan concedes that if you view the spectacular subjectively it can be horrifying. 'So many animals die, and I can imagine people wondering why we aren't doing something about it – these are living creatures that enjoy human sanctuary all along the migration route, yet here they die.'

continued

This mutual dependence amongst males appears to have bred a 'gentleman's agreement' concerning the females and mating. They rarely fight amongst themselves and the first to find a female on heat is usually accorded a dominant position over the other males.

What was once regarded as peculiar aspects of take-over behaviour – the killing of some of the cubs – has now been revealed through long-term field studies as adaptive: the males do not want to waste time and energy rearing others males' offspring, so they kill them. Also, the females may then come on heat more quickly and be fertile for the new male.

JONATHAN SCOTT
Maasai wildlife

Below: *The greatest wildlife spectacular on Earth – or the greatest tragedy? The bridge of death which forms at the Mara river when hundreds of thousands of animals, mostly wildebeest, strive to reach the paradise of the lush red wheat grass plains beyond.*

An insoluble problem?

He has neatly pin-pointed the dilemma of nature-watching. Nature in the raw can often be very raw indeed and it requires a harder heart than most of us possess to be able to stand back and see all this as a part of 'life's rich pageant'.

But that is exactly what it is, and we can do damage if we interfere. Ironically, the very existence of the Serengeti and Mara sanctuaries – both fine examples of the better side of human nature – have made the number of animals dying at the river even worse.

In 20 years the wildebeeste population has expanded from 300,000 to the present figure of approximately two million. The river crossing and the other hazards of the migration are a natural cull. Without them there would undoubtedly be calls for unnatural culls. Experience has shown that once a reason is found for shooting wildlife, people grow dependent upon this new resource and will not give it up easily.

Where possible, it is best to let nature take care of its own, and as Jonathan is quick to point out: 'They are not just dead bodies, but essential food for the other creatures that fit into this unique pattern of wildlife. As well as all the scavengers and the carnivores, there are all the smaller creatures: a huge realm of species directly dependent on this situation.

'Nothing is ever a waste. We can't afford to interfere at all. It's just too big for that. It has to be allowed to progress just as it is, the natural way.'

The future

That leaves just one question which, unfortunately, we always feel obliged to ask: will the Mara survive?

Jonathan Scott has a thread of doubt. A large area of very similar grassland outside the Mara sanctuary (and still well to the north) has recently been planted as a wheat prairie and the yields are good and profitable. The Mara could become another Kansas, and the economic pressure, given Kenya's very high birth rate, is there. Yet Jonathan believes it will survive: 'Just look at it,' he said with great confidence. 'Its beauty is unparalleled. If we can't manage to save an area like this, we won't save anything worth having.'

We obviously share his prognosis, but that wheat prairie to the north is perhaps more ominous than he realizes. It means that the once nomadic, cattle-herding Maasai, who own the land, are rapidly changing into yet another group in competition with the wildlife and its habitats. We devote pages 224–233 to a consideration of that grave prospect with a Maasai conservationist, Solomon ole Saibull.

JONATHAN SCOTT
Maasai wildlife

9. BIG GAME-WATCH

DAPHNE SHELDRICK:
There are no wild animals

Daphne Sheldrick was born in Kenya and has been married to two of the country's legendary game wardens, David Sheldrick and Bill Woodley. Her brother, Peter Jenkins, is another famous warden. She is an amateur nature-watcher.

When we first met Daphne Sheldrick in 1982 she was already internationally renowned as an animal-lover and nature-watcher extraordinary; this was a result of the film *Bloody Ivory* which centred on the work of her husband David and the war he fought against elephant poaching in the Tsavo National Park in the late 1960s.

Daphne then ran an orphanage for the creatures left to die by the poachers: baby elephants, rhino, antelope, warthogs, and birds – a whole menagerie of animals. All these creatures walked about free at Tsavo, and were often to be found lolling about on her front lawn. Rhino allowed her to give them mud baths, zebras became firm friends with guinea fowl, full-grown elephants let Daphne feed their babies under the mothers' bellies because it made the infants more secure, and her hand-reared antelope came home every morning after spending the night with their wild relatives! For Daphne there are no 'wild animals', only creatures living free.

Since *Bloody Ivory* was filmed nearly a decade ago, almost everything has changed for Daphne. David died most tragically of a heart attack just when it seemed he was winning his private animal war. In a generous act of understanding the Parks Department gave Daphne, who has never lived away from the embrace of the wild, a building plot overlooking the Nairobi National Park just outside Nairobi. Here she now writes educational articles on wildlife and administers the David Sheldrick Memorial Trust which promotes practical conservation (mainly in Africa).

As far as Daphne's attitude to animals is concerned, nothing has changed. On our way out to her house in a taxi, the driver assumed we were tourists and gave us a cheerful monologue on the terrors of the animals we might see. When we drew up outside Daphne's

Above: *Daphne Sheldrick has always shared her homes in Kenya with animals others regard as strange. Today she has 'wild' wart hogs as watchdogs.*

Below: *Her home, near Nairobi, Kenya.*

Opposite: *Daphne Sheldrick nursing a baby elephant.*

DAPHNE SHELDRICK
Big game

bungalow, however, his eyes bulged and he refused to remove our luggage.

His gaze was fixed in horror on four dusty creatures browsing contentedly in a plastic bowl of chicken feed, a few feet from Daphne's kitchen door. 'Dem's wert hugs,' he hissed. One of the wart hogs looked up and obligingly waggled its sharp tusks to confirm this identification.

Daphne appeared at the doorway, and noting the driver's concern called out cheerfully: 'Don't mind the piggies.'

This was the first time we had actually met this remarkable woman. We had, however, recently seen a pair of wart hogs in the Maasai Mara seeing off two attacking lionesses.

'Are they tame, Daphne?' By now she was standing by the car watching the hogs with some pride. 'No, they're not; they come in from the park. They're my watchdogs. But don't worry, they won't bother you.'

During the next few days we learnt that she was right. Halfway through dinner that evening, Daphne suddenly squealed with excitement, snapped on an outside light and drew back the curtains to reveal a full-grown rhino sedately crossing the lawn. This gentle giant blinked in the light, cast Daphne a mildly disapproving glance, and continued on to the small water hole she has dug at the bottom of this new garden. Daphne told us: 'I also put out salt for them. They don't get nearly enough in the wild.'

Every morning she takes scraps out for her chickens and for a huge African kite which swoops down to just a few feet above her head to take the morsels on the wing. About once a week Daphne is woken up by the lions coming into the garden to take a smaller creature at the water hole. Some, like one of her bigger wart hogs, survive with minor wounds, and Daphne feeds them during their convalescence. She is just as proud of her hunting lions as of her wart hogs; it means that life in the garden is healthy and normal.

On our way to bed in the little rondavel in the garden that she keeps for guests, Daphne told us to check that the draught-proofing was in place round the door. With the temperature in the high twenties (degrees Celsius) we would have welcomed the air, but Daphne smiled: 'It's not for draughts; there's a nest of spitting cobras in the kit-store and the little ones have taken a liking to the warmth of the rondavel.' But like all the creatures under her wing, the baby spitting cobras left us alone.

Tsavo: stepping back into the past

A couple of days later we flew with Daphne to her old home; Tsavo is rather a sad place today. Although poaching is nothing like the shambles David Sheldrick fought a decade ago, neither the park nor the animals have fully recovered. It is a huge, awesome kingdom the size of Israel; remote, arid, sparsely vegetated and rightly famous for its truly wild animals. The lions are ferocious and scarred. The elephants were famous for their huge tusks, hence their attraction for poachers. The park was regarded as the stronghold for black rhino in Africa. 'David's diaries have entries like: "Twenty-four rhino seen in a drive down the river," which was quite commonplace. In some places it wasn't safe to walk on foot because there would be rhino boiling out of every bush,' Daphne recalls.

Tsavo experienced almost a decade of intense poaching from the mid sixties. These days you can drive all day without even seeing one of the few remaining large herds of elephants. We experienced this sobering reality ourselves. As for rhino, this is the grand tragedy of Tsavo. 'You could wander around Tsavo for three months and not even see any tracks,' Daphne said sadly.

The endangered rhino

In the course of our several journeys to East Africa in recent years, the plight of the black rhino has progressively worsened to the extent that no one knows quite how many are left. While we were in Nairobi in 1982, a senior politician announced that the rhino population for the whole of Kenya might have fallen to 100.

The political furore this announcement caused resulted in a hasty retraction, and recent reports indicate it may have been unduly pessimistic. We believe that East African conservationists are understating the number of rhino still alive in remote areas in the hope that poachers will leave them alone.

In Tsavo itself the scarcity is a reality, and all Daphne can do today is to tell stories of the plentiful past, in support of the movement to remove a breeding stock to rhino 'stud-farms'.

Daphne is already involved in the most ambitious of these projects, as she is particularly well-equipped to advise on the raising of young animals and their re-introduction to the wild. She did exactly that for years with the animal orphans of Tsavo.

'I'm particularly fond of rhino,' she said. 'I think

DAPHNE SHELDRICK
Big game

they are misunderstood. They're always depicted as aggressive, constantly bad-tempered and charging. In fact they are not like that at all. I found rhino one of the very easiest animals to tame. You can tame a full-grown adult in two days.'

The picture below supports this statement. Over the years, Daphne has raised several rhino of all

Above: *Two orphans and a story of unrequited love! Punda the zebra became so fond of Stroppy the rhino, that it tried to mix with wild rhino when Daphne Sheldrick took her menagerie for walks. 'We were forever patching him up,' Daphne recalls. 'But Punda had obviously decided to be a rhino.'*

ages, including the tough little creature Stroppy, pictured here with the love of its life – a zebra called Punda. Punda came to the orphanage when he mistook a striped minibus full of tourists for his mother and followed it back to one of the Tsavo lodges. The baby rhino was a victim of poaching, and refused all food in the compound until the little zebra arrived.

'Punda marched straight up to Stroppy and started chewing his horn.' Unaccustomed to such disrespectful treatment Stroppy seemed to pull himself together: 'From that moment on they became firm friends and grew up together. They were inseparable.' Daphne became convinced that the zebra had come to think of itself as a rhino. 'Punda never learnt to differentiate between tame and wild rhino. As a result of these encounters we periodically had to patch Punda up.'

Telepathic antelope

Similar relationships, which in any other context would be regarded as highly unlikely, proliferated under Daphne's tender loving care. Ferocious 'wert hugs' shared the front lawn with hand-reared antelope for afternoon tea.

'The antelope generally regarded them as an absolute pest. The wart hogs hated to see anyone relaxing and would snuffle around until the impala eventually had to move. However, we had a young kudu, Jimmy, who was made of much sterner stuff. He just refused to move even when the big-tusked adult wart hogs tried to nudge him out of the way.'

Daphne developed a real empathy with the many antelope she cared for and raised over the years, in particular with the impala: they became so accustomed to her that they even allowed her to watch them give birth.

From such intimate studies, and from their behaviour in the bush, Daphne came to some very unusual conclusions about antelope. 'I'm absolutely convinced that they have telepathic abilities,' she confided to us. 'I noticed it first with Bunty who raised several of her young under our wing. These animals shared their time between us and the wild herds.

'I used to watch Bunty staring into the distance for long periods when her offspring were away. She sometimes stood there for an hour or more, as if she were concentrating on something, although there was nothing visible.

'Then, several hours later, one of her offspring would appear. This happened quite regularly so that when we saw Bunty behaving like this we could actually predict that one of her sons would return that day. We were never wrong.'

Memorable elephants

Most of this conversation took place on the steps of a makeshift house in Tsavo which is the home of one of Africa's leading wildlife cameramen, Simon Trevor, who filmed *Bloody Ivory*. Here, one hot morning, Daphne demonstrated how it is possible to achieve a lasting friendship with a wild animal.

DAPHNE SHELDRICK
Big game

Below: *A strange meeting of old friends. Though they now live many miles apart, Daphne Sheldrick is still recognized with affection by the two elephants she raised from tiny calves after they had been orphaned by poachers.*

Out of the red horizon walked two huge elephants. With a smile of delight, Daphne went to the rail and called: 'Eleanor . . . Eleanor!'. The larger of the two beasts, a very large elephant indeed, performed a distinctive elephant waltz across the yard and wrapped its trunk affectionately round Daphne's neck, and then round Julian's!

We immediately realized that this was the famous Eleanor, raised by Daphne, and the star of the orphanage sequences in *Bloody Ivory*; here she was, still alive, well and living happily in Tsavo.

Next the equally famous young male, Bukanesi, entered the scrubby garden. He was raised by Daphne and Eleanor, working together as a team. Bukanesi's mother was poached and the calf stood for three days beside the body of its dead mother. When David Sheldrick brought the calf home (Bukanesi means 'the weak one' in the local dialect), no one thought it had a chance because at that time nobody had managed to

Left: *The elephant, Eleanor, played surrogate mother for Daphne to a number of foundlings, including, unlikely as it may seem, the baby rhino, Stroppy. Here the two share a hose-pipe, Eleanor having learnt how to turn on the garden tap!*

wean a young elephant which was still reliant on its mother's milk.

Eleanor adopted the calf and gave it the security of a parent, while Daphne battled night and day to find the magic formula for the milk. 'Elephant's milk is quite unlike the milk of any other animal. They can't assimilate the fat in cow's milk, yet they do need a great deal of fat. So we had to start with a fat-free base, then add something. We had tried almost everything until I thought of coconut milk, plus lots of vitamins and other things.

Finding the right formula was only the first obstacle. After that, it was a matter of infinite patience and sheer hard work. 'You feed a rhino every three hours, but not at night. You feed an elephant at night too, so after a year you are walking around like a zombie,' she laughed.

But the patience and the formula, and most of all

DAPHNE SHELDRICK
Big game

the teamwork with Eleanor, worked. When Bukanesi refused to take milk from Daphne's gallon bottle, Eleanor allowed herself to be used as a surrogate mother. Daphne used to feed the infant beneath Eleanor's huge belly, fooling the baby into believing it was being breast-fed.

Eleanor, her trunk wrapped affectionately round Daphne, was now 22 years old, and Bukanesi a large, healthy, young male. Yet both are still wild creatures. Admittedly they enjoy special protection in Tsavo, and are guarded day and night by their own squad of game guards. They may wander as they please, and Daphne's relationship with them continues even though her visits to Tsavo are now few and far between. When they greeted her with familiar affection on this occasion, she had not visited them for a year! While we found this remarkable, it came as no surprise to Daphne, because of her particular attitude to animals.

Communication and caring in wild animals

Daphne Sheldrick believes that all animals communicate much more effectively than we give them credit for, or are even aware of (impala telepathy is a case in point) and, to a greater or lesser degree, she believes they also care selflessly. She totally rejects the suggestion that animal altruism is just herd instinct.

We had observed that when Eleanor and Bukanesi finally left Simon Trevor's somewhat trampled garden (Eleanor having turned on the garden tap and sluiced down some 70 gallons of water), they both paused to fondle a pair of elephant tusks lying on a rock near the house.

'They have a curious fascination for their dead,' Daphne commented: 'They seem to understand that their tusks have been the cause of their undoing. I once saw Eleanor try to pull the tusks out of an elephant's skull when, to our knowledge, she had never seen a dead elephant before. She concentrated on the tusks, working them slowly up and down until they were free. In the wild you often see the tusks of dead elephants that have been shattered against rocks and thrown away.'

Daphne is a mine of information on elephants that have displayed altruistic features such as compassion and grief. 'It was a long time before we were able to raise the very young calves,' she recalled. 'Whenever one of them died, Eleanor was desperately upset. She would protect and care for any member of her herd, to the point of chasing lions away. I do believe she cared very deeply for their safety. The fact that her "herd" contained creatures other than elephants didn't matter, nor does it matter for wild elephants. I know of an instance where a pair of elephants assisted an aging buffalo through the last few days of its life, supporting and protecting it.'

We questioned the validity of the elephant as a general example of an animal's intelligence, pointing out that even scientists regard elephants as particularly intelligent, on a level with the advanced apes and the dolphins. Daphne would have none of this.

'It is very difficult to gauge the intelligence of animals: we usually try to measure it by comparing it with our own intelligence, but we don't appreciate the fact that all animals are geared to different systems and their intelligence must therefore take a different form to ours. This is one of the arguments I have with a lot of my scientist friends: that I shouldn't attribute human feelings to my animals. I argue that perhaps I know the animals a little better than they do and that animals *do* have the same sort of feelings as humans. Some are good-natured, some have a sense of humour, while others can be bad-tempered, long suffering or impatient.'

Daphne then let us into a secret: 'If you are really fond of animals, and have a deep empathy with them, they will respond to you. There's no good in pretending that you're fond of them, because they will know instinctively whether you are or not. This is the secret of handling animals.'

DAPHNE SHELDRICK
Big game

10. MONGOOSE-WATCH

Not all the animals which attract nature-watchers to the wide plains of East Africa are 'big' game. (The term, incidentally, is inherited from the days of hunting safaris when Europeans who wanted a change from shooting their 'small' game such as rabbits, hares, pheasants, grouse, came out to Africa to slaughter the big game.)

ANNE RASA:
The intricacies of the dwarf mongoose

Just down the road from Daphne Sheldrick's old home at Tsavo a young Welsh woman, Dr Anne Rasa, has been persuading a German University to support what may be the most protracted field study ever. The object of this attention is, at first glance, one of the least spectacular animals in the bush: the southern dwarf or pygmy mongoose. It has taken Anne Rasa a decade to reveal just how remarkable this little mammal is.

There are parts of Kenya which could be called the Garden of Eden: lush, heavily forested and rich with plants. Anne Rasa's hideaway is not one of these. It is a measure of her commitment, and an interesting pointer to the cocoon which nature-watchers sometimes weave around themselves, that she still finds her arid landscape 'the sublime experience, the Nirvana'. She lives 45 miles from the nearest town and, apart from two trusted camp attendants, she is alone in a tent for the three months (January to March) of field work she puts in each year. While she is there she spends 13 hours a day observing and then about two hours writing up her notes. Once she has started with a group of mongooses she stays with them for at least 28 days.

It can be very hot, sometimes the temperature reaches 120°F. Everything for miles around is wild and potentially dangerous, even the cattle. The only well-documented account of a man-eating African lion concerned one of the very large local lions. However, Anne is not particularly worried by them. 'Since I've lived out here my senses have become extremely acute and I've learned to recognize dangerous sounds, although it's sometimes very difficult to get to sleep when the lions are roaring at each

Opposite: *Southern dwarf mongooses on a termite mound – the most caring colony?*

Above: *Anne Rasa has spent more than a decade in the hot-spot of Tsavo, Kenya, studying one of the smallest, yet most active of African animals, the dwarf or pygmy mongoose.* Below: *Tsavo, Kenya.*

ANNE RASA
Mongooses

MONGOOSES

The mongooses are members of one of the oldest and most diverse carnivore families. Fossil records of their ancestors found in Eocene stratas (40–50 million years ago) reveal features, including the basic skeleton, that have hardly altered since then.

The *Viverridae* comprise 66 species spread among 37 genera of which the best known are the civets, genets and mongooses (subfamily *Herpestinae*). They all tend to be omnivorous, eating insects, small mammals, birds and eggs. The family is distributed in Africa south of the Sahara, Madagascar, the Near East, Arabia to India, Southern China, Southeast Asia to Borneo and the Phillipines, and southwest Europe. It has been introduced in the West Indies, Fiji and the Hawaiian islands.

The Southern Dwarf Mongoose (*Helogale parvula*) is the smallest member of the 31-strong subfamily of *Herpestinae*, with a body of some 25 cm, weighing about 30 kg. They all live in Africa, from Ethiopia to northern South Africa, west to Angola, Namibia and Cameroon. Their coats vary from dark brown to light grey with a 'grizzled', speckled overlay.

Dwarf mongooses are the most social of the *Viverridae* (few other mongooses are social at all) and often live ten years or more, twice as long as others of their kind. The average size of a pack is 8–10. Females sometimes remain in their natal group but most eventually migrate to other packs or join up with other emigrants to form new packs. Each pack is dominated by a breeding pair and, because they are monogamous, a male must normally attain 'alpha' status in order to breed and raise offspring.

The evolution of the strong social bond governing these mongoose societies is thought to have been promoted by predation, given that the animals find their food (mostly invertebrates) individually as opposed to other social cats (e.g. lions, hyenas, wolves), on whom the important selective pressure to sociality has been the success of hunting in packs rather than alone.

other only five yards away from my tent. In my mind I know I'm in no real danger. The only animals I'm really afraid of are the elephants and bees, because you can't get away from them in a truck. There's a waterhole nearby where 20 to 30 elephants often go to drink, grumbling, shrieking and moaning. If one doesn't like you, it can throw your truck over and kneel on it. There is also a very aggressive African bee with which I've had a couple of nasty brushes. If they start after you there's nothing very much you can do but put your foot on the accelerator and get out as fast as you possibly can.'

The place is also stiff with venomous creatures such as mambas, puff-adders, scorpions, centipedes and even bird-killing spiders. Nirvana? Perhaps Dr Rasa has lived there alone too long.

But, of course, she is not alone: those 14 years have been spent with a group of animals which her extended studies have revealed may be socially closer to us than the great apes. Dr Rasa is a student of another great nature-watcher, Professor Konrad Lorenz, whom she quoted when asked to explain her obsession. 'Lorenz once told me that objectivity is all right, but before you can be objective you have to be subjective, and the only way to be subjective is to *be* your animal.'

We have heard this from Lorenz himself and we believe it to be the key to good nature-watching, amateur or professional, even though it appears to contradict one of the first rules of ethology, that all study should be detached.

It is almost impossible, of course, to remain detached from creatures as appealing as dwarf mongooses. They occupy territories of 0.5–0.8 square miles and they do a circuit of this every 26–28 days. These are Anne Rasa's figures: one of the advantages of long-term field studies is detail and accuracy.

A day in the life of a dwarf mongoose
Mongooses get up at about 6.30 a.m., having sent up the 'first guard' to check that the surroundings are clear of predators, particularly the pale chanting goshawk. Once the group is assured that all is safe, they set off. It is a matriarchal society and so the highest-ranking female, the 'alpha' female, decides where they will forage that day (for insects, lizards, snakes, small birds, eggs and fruit).

'I can tell whether they are going on a long or short journey by how early they get up and how long they spend grooming on the termite mound where they have spent the night. They usually have about 10 to 15 minutes of social activity on the mound; they mark, everybody grooms everybody else and they deposit their faeces in a little toilet or "midden", off to one side. They are extremely clean animals.'

Eventually the mongooses set off on their daily tour, carrying their young. Anne Rasa can recognize all the members of the Tsavo group and she quickly identified its most important member, the alpha female, when that creature emerged. 'She's the group leader,' Anne reminded us. 'She has a lifelong monogamous relationship with the alpha male and is the only female allowed to raise young.'

At the end of the day's foraging the family finds a suitable termite mound in which to spend yet another night. Termite mounds are ideal for the little mongooses that live in the ventilation shafts where the air is 'air-conditioned' to a temperature both ants and mongooses appear to prefer (even though it is 93°F).

Aggressive behaviour

'The alpha male leads the serious fighting. There are two kinds of male fighting groups, one I call the shock troops, the other the frontline troops.

'The frontline troops are made up of the young animals, up to three or four years old. They are used in situations which are not intrinsically dangerous, such as when their group meets another group. The frontliners form into an amazing battle array – a wedge formation with the alpha male (their father) in front, the yearlings behind him, then the three to four-year-old subadults, with the strong adult males at the back, and the alpha female with the very young right at the rear.

'They have terrific fights. My last study showed these yearlings to be the most aggressive (but not strongest) members of the group. At first I simply couldn't work out why they were putting their weakest members in the front line. In fact when they start fighting the yearlings go in with a tremendous clash and can get chased up to a mile away. Quite often they can't find their way back to their own group. Instead, they trail another group and sleep in an adjoining mound. Slowly and by continued proximity over a period of five to six weeks they're allowed in, and so the group receives an injection of new blood.'

When a potentially lethal situation occurs, however, the mongooses adopt completely different tactics, and send in the shock troops.

ANNE RASA
Mongooses

Below: *Getting their act together: it appears from Anne Rasa's observations that hornbills wake up tardy mongooses (asleep in the termite mound) so that they can share the insect bounty of the mongooses' busy foraging.*

'In this situation the line-up is reversed. The adult subordinate males come right to the front, the alpha males goes off with his wife and makes big threatening movements but doesn't jump in until it is really necessary. Behind the adults are the subadults with the yearlings right at the back. So when it's possible that an animal could be killed in an interaction the strongest members of the group are at the front.'

Mongooses and hornbills: reciprocal benefits
One of Anne's most remarkable discoveries concerns a mutualistic relationship between the mongooses and two hornbill species: the yellow-billed hornbill and the decan.

'They work together for the mutual good of each other's communities. It's the only case of mutualism to be recorded in two so highly developed vertebrates. The mongooses use the hornbills as an early-warning system for birds of prey, while the hornbills use the mongooses as beaters for leaping and flying insects,

especially grasshoppers, which they love to eat.

'I've even seen the hornbills wake up the mongooses when they have been late getting up! The birds fly down onto the mound, stick their heads in the holes and go "wak, wak, wak". They also do something I call "chivvying", which is rather amusing. This is very anthropomorphic, but if the hornbills are really hungry and the mongooses' social time on the mound seems to be going on too long, the circling hornbills land, walk very purposefully off into the bush, then stop and look back.

'If the mongooses don't follow them they'll repeat the performance. I've never seen them do it more than five times before the mongooses follow them off into the grass.

'The mongooses' social time on the mound usually lasts for about 15 minutes, but if they're woken up by the hornbills they never take longer than three minutes.'

Anne Rasa is aware that she has discovered an extraordinarily advanced behaviour, but even she is not quite sure of its boundaries: new features come to light every time she studies the phenomenon.

'Sometimes the hornbills don't turn up. If there's a bird of prey around, they move away into the bush. When this happens, the mongooses wait for them; as soon as the hornbills return, the mongooses are off like a shot. Amazingly they have taught the hornbills who their enemies are, even creatures that aren't enemies of hornbills.

'When a mongoose predator is in the vicinity, the hornbills take off in an explosive vertical flight into the trees. If it's a reptile they give what I call a "warn wak" call – a very rhythmic, slow "wak, wak, wak" which carries for a great distance. A hornbill making that sound half a mile away will cause the hornbills nearest the mongooses to fly up. As a result the mongooses have an aerial guard extending round the mound with a mile circumference.'

For their part, the hornbills are rewarded with insects that the mongooses flush out of the grass by their busy passage round the territory. Mongooses don't run after these agile prey because it would make their presence obvious to predators. Instead the hornbills claim these prizes.

Mongooses and reptiles
Anne has spotted other aspects of mongoose behaviour which verge on mutualism. 'Mongooses have a strange

ANNE RASA
Mongooses

147

ANNE RASA
Mongooses

relationship with the big lizards, which also live in termite mounds, and many actually share them with the mongooses. The lizards eat their faeces, from which they derive considerable moisture and protein. The baby mongooses play with the lizards even though they are twice their size, and steal food from them.'

This near-mutuality does not apply, however, to all the reptiles. Snakes, for example, are definite enemies, although Anne has shown that the antipathy between snakes and mongooses immortalized by Rudyard Kipling in his *Rikki-tikki-tavi* stories is grossly exaggerated and has, in fact, caused a wildlife tragedy.

'If Kipling hadn't written *Rikki-tikki-tavi* and the idea had not got around that mongooses eat snakes, there would still be birds on most of the Caribbean islands and on Hawaii. Having read the stories, the inhabitants of these islands introduced mongooses to the islands to kill the snakes. The species they chose, the Indian mongoose, is primarily an egg-eater, so now practically all the indigenous birds on those islands have been eradicated or brought to the edge of extinction.

'In fact, snakes are not food for mongooses, they are simply enemies. Mongooses differentiate between species according to which are enemies and which are not. They have a special call – the ''come quickly call'' – on which they all run to have a look. They only do this when they see snakes. If it's one of the non-predatory snakes, the male mongoose will threaten, but once they've all had a look, they wander off and leave the snake alone.

'If it's a predatory snake (there are two species of cobra found in the region, two mambas, and a puff adder) the response of the mongoose is quite different. They surround the snake and immediately start to attack it. These snakes are terrified of mongooses and try to get away as fast as they can; in most cases the mongooses just chase them off.

'If they do kill, it's with a bite in the neck. One mongoose attracts the snake's attention, making it strike. As the snake recoils and before it can strike again, the mongooses attack and bite it three inches behind the head. I measured it exactly, but it didn't occur to me until I tried to pick up a snake by the back of the head, that three inches from the head is the exact fulcrum of the coiling action: by going for that point first the mongoose paralyses the snake before it can coil around. Then it kills with a second bite across the eyes.'

(One last intriguing observation: Anne Rasa has discovered that the mongooses are immune to the venom of the snakes which share their habitat.)

Talking to mongooses

If Anne Rasa becomes lonely she can always listen to the mongooses talking to each other. After 14 years, Anne has learnt to recognize, and even talk like a little mongoose.

'I think the number of pure sounds they make is probably unrivalled among the ''lower animals''. They have 18 different vocalizations which, when arranged in different ways, make up varying mixtures. Now when I hear them talking I know exactly what they're saying and who's saying it.'

She then announced that she had some evidence for mongooses relaying abstract information. 'It's always said that Man is the only mammal able to give

Below: Long-term field studies often destroy much-loved theories. Anne Rasa has shown that mongooses don't eat snakes (like this puff adder). They are simply enemies to be killed with a swift bite to the neck, or, if the snake is not predatory, to be ignored.

ANNE RASA
Mongooses

abstract information. There is, however, some evidence of this in chimpanzees and I believe I have evidence for it in mongooses. In one case it was very clear that an abstract item of information had been given and was acted upon 12 hours later.'

The mongooses in question were a young male mongoose, Rusty, and his brother, Goldie. Rusty was alone on a mound one afternoon when a black-tip mongoose, living in a nearby mound with its group, tried a surprise attack. Brother Goldie gave Rusty a sound warning, then teamed up with him to drive the black-tip off. 'There was no touching between mongooses from different groups to cause smell contamination,' Anne stressed. 'Immediately afterwards the two young mongooses went down into their own mound for the night.

'The following morning, however, the mongooses got up very early and there was no social activity on the mound. Instead, the alpha female gave the moving-out call, which gathers the whole group together, and they all made a beeline for the black-tip's mound. The adult subordinate males were sent in first; they went down into the mound, their coats "bottle-brushed" and obviously ready for action. After a while it became evident that the black-tips weren't there so the mongooses came up and left a huge pile of faeces round the entrance which the black-tip had used.'

This would certainly seem to be a case of the use of abstract information, which has never been entirely proven in any group of animals, but as Dr Rasa points out it also raises another intriguing question which could set the communication abilities of mongooses in a new, intriguing light.

'How did Rusty and Goldie tell their mother that the black-tip had tried to beat them up the previous night?'

Anne Rasa believes that the behaviour of mongooses definitely casts a new light on the 'intelligence' of mammals, and she has other examples to support the hypothesis. For instance, she has proved that they have excellent memories. 'In 1976 I started feeding and taming a group living near one of the Tsavo lodges,' she explained. She wanted to make an intimate study of certain aspects of behaviour, including their methods of dealing with snakes and other predators.

'During the week I was there they got so used to me they would come when I called. Then I went away for 18 months, assuming that when I returned I would have to start all over again. In fact, when I arrived and called them, their little heads popped up and they all came galloping towards me. The people at the lodge were flabbergasted. They had not come to anyone else in all that time. Since then I've visited them about once a year, always with the same results.'

Altruism

But perhaps the most intriguing of Anne's discoveries is what we call 'altruism' in the mongoose family.

Altruism (unselfish concern for the welfare of others) is a form of behaviour we generally assign only to ourselves. Obviously, the example of Rusty and the predatory black-tip mongoose is a story of caring, but Anne Rasa has detected better examples. 'Mongooses are the only mammals where scientific documentation shows that they take care of their sick. There are many hunters' tales about elephants helping to carry off their wounded, but mongooses will stay with their sick until the victim either dies or recovers.'

Three years ago Dr Rasa was observing a subadult male mongoose foraging in some bushes when she heard the 'come quickly' call. The rest of the group rushed towards a certain bush and started attacking something which she correctly assumed to be a snake. Eventually the group quit and bedded down in a nearby mound.

continued

inheritance of an altruistic gene is at least theoretically possible.

The theory admittedly begs a lot of questions, not least its apparent absence in closely-related groups who should, seemingly, have preserved a gene that could be beneficial. Long-term field studies of such groups (e.g. chimps and mongooses) are, however, starting to produce evidence which could extend the altruistic characteristic outside the exclusive domain of man.

The American anthropologist, Colin Turnbull, believes that it is not a genetic characteristic at all, but an emergent property of animal societies: 'our much-vaunted human values are not inherent in humanity at all, but are associated only with a particular form of survival that we call society, and that all, even society itself, can be dispensed with.' (*Thr Mountain People*, 1972). Altruism is almost certainly an aspect of nature about which we still have a great deal to learn, and there is every reason to believe it is not entirely a matter of human nature.

ANNE RASA
Mongooses

Below: *Eggs form an important feature of mongoose diet. The animals crack them by kicking the eggs against rocks with their front feet – rather like a ball from a rugby scrum.*

'The next morning everything started strangely. Instead of getting up and going to forage, the mongooses stayed at home all day and were just "scatter foraging" in circles close to the mound.

'The next day, however, the same subadult male came out and I realized that the snake must have bitten him in the belly. It was probably a sand boa: they usually lie half-buried, and this one had struck the mongoose in the lower abdomen, ripping off all the skin and crippling him. He couldn't even put his leg on the ground!

'When one of their number is sick, the mongooses cuddle up to him – literally pile on top of him – and groom. The alpha mother takes a particularly active role in this, which is unusual because she doesn't normally bother much with the babies.

'For the next five days these mongooses (there were 12 in the group) shuttled between two mounds that were just 11 yards apart. The sick one would try to hobble with them. They caught food and fed it, which

they never usually do. On the fifth day, when it could almost walk normally, they went back to their usual foraging.'

It seems that many of the altruistic behaviour patterns the mongooses demonstrate are learned during an individual's lifetime, rather than being 'innate' (inherited in the genetic make-up) – an even more remarkable phenomenon in so 'lowly' a creature.

'One of the most important patterns for the survival of the group (itself altruistic) is guarding,' Anne continued. 'An animal will climb to a high place and then usually guard in a direction 180° away from the group foraging, thus operating as a rear guard. This behaviour is actively taught to the young animals. When they are about five months old they try to climb up and sit with the guard for a while. They have a receptive phase for this behaviour. The guard will teach what is dangerous and what to look out for. The guard leaves the baby alone for a while. The baby will watch, then realize it's alone and start to give the "where are you" call. Then the guard will come back. These periods of leaving the baby alone are gradually extended until it fits into the system. This is a long process that can take up to two years before the baby is an efficient guard.'

Anne Rasa suggests that this makes sense of altruistic behaviour. It is well worth the mongoose group investing a week or two in saving one of their number which is an effective guard, compared to the time it would take to train a new guard.

We will close as reluctantly as we left Anne's Nirvana-like camp in the Tsavo wilderness, with her most intriguing claim which is a springboard for the subject we will consider next, the emergence of primate life.

'I think the dwarf mongoose society works in the same way as the ideal human society did, and still does for some of the very primitive people who live in hard conditions,' she proposed. 'It is a wholly altruistic society. Everybody helps everybody else because if they don't, they die. Mongooses have the highest level of altruism known for any animal except Man.'

11. LEMUR-WATCH

How did we acquire the special tendencies towards altruism? It is generally accepted that they are a product of our social background and may well date from the time when primates first formed social groups.

The first modern primates to evolve were the lemurs. Their descendants, the fugitive nocturnal possums and bush babies, are alive in Africa, but only on Madagascar have the lemurs themselves survived. The island drifted away from mainland Africa during the Eocene era some 60 million years ago and avoided colonization by monkeys. Thus the Madagascan lemurs were able to exploit the habitat niches that were taken over by monkeys elsewhere.

The island of Madagascar is 1000 miles long, closer to a small continent than an island. Of the trees that once cloaked the entire island only a ring of coastal forest now remains. Two-thirds of all the land has been slashed and burned for primitive agriculture resulting in terrible erosion of the soil. Amidst this wasteland a scrap of forest, the Berenty Reserve, has been maintained as a sanctuary for the lemurs by the de Heaulmes family for more than 30 years. So Berenty is a living museum where nature-watchers can study these primates and reflect upon the significance of their findings for human behaviour.

It is possible to see as many as ten different species of lemur in a single Madagascan forest. This includes the little lemurs which, at a mere three ounces, are the smallest living primates; the dwarf lemur and the hairy-eared dwarf lemur, and a group known as the true lemurs which are cat-sized quadrupeds weighing about seven pounds. Probably best known is the ring-tailed lemur, a beautiful creature with a pelt of pale grey and white, a tail of black and white bands and coal-black patches distinguishing the eyes.

ALISON JOLLY:
The oldest families

Dr Alison Jolly, the American nature-watcher, has been finding every opportunity to visit the Berenty Reserve to study the behaviour of lemurs since she discovered it in 1962. We were fortunate to be able to join her on her field trip in 1984 when she was gather-

Above: *Alison Jolly has been studying lemurs on the island-continent of Madagascar, off the coast of Africa, since 1962.*

Below: *The island of Madagascar.*

Opposite: *A sifaka lemur – laying down the ground rules of the human family?*

ALISON JOLLY
Lemurs in Madagascar

Right: *Ring-tailed lemurs, with baby in the travel position.*

ing more information on the social behaviour of these creatures. She recalled that her studies were sparked off by asking herself the following question: 'How much of human love dates back to this family of pri-mates that is far older than humans? They care for each other and they too live social lives.'

Ring-tailed lemurs

She first drew our attention to a group of ring-tails. The most distinctive feature of this beautiful lemur is that, unlike other lemurs, it is semi-terrestrial: groups of ring-tails spend much of their time on the ground patrolling their territories. A group is made up of 15–25 animals including several adult males. Males mark their territories using a large glandular patch on each wrist, which they gouge deeply into tree branches with a horny spur. (They also have scent glands near their armpits.) The long tails, held like graceful question marks, are raised and 'shivered' at opposing males during their most aggressive territorial displays which are known as 'stink-fights'.

The ring-tails have also developed sophisticated verbal communication in addition to their complex signalling: males sing between troops with a high wolf-like call; the troop uses a high-pitched bark in chorus to deter or warn each other of a ground preda-tor, while their main flying predators, the eagles and

hawks, are heralded with screams. In close contact, members of the troop 'talk' to each other with cat-like purrs, and clucking and mewing sounds, as they move rapidly through the forest. 'I'm trying to find out why there are so many loose males. You see these are very seasonal animals; there's a two-week orgy of mating in late April and almost all the babies are born during two weeks in September.'

We were there in November, by which time most of the males had changed troops. 'First they wander further and further from their own troop,' said Dr Jolly. 'I would particularly like to know if these changes, which can affect a quarter of all the males, tie in with the differences between lemur society and monkey society. It is only here that lemurs live in families and troops, the First Family if you like.'

Lepilemurs

We went out at night to observe the lepilemurs which inhabit the spiny desert part of the reserve.

'These look very like one of the original arrivals who floated here on rafts of tangled branches, before Africa got too far away,' Alison pointed out. 'We have found fossils in Europe and North America which look very like lepilemurs.

'Their social system is a harem but the members of this group rarely associate – they're not a very tight family. Big males may tolerate much smaller ones, but not real rivals. Females may share the range with their growing daughters, but only for a while. The big males also know who their females are and roughly where they are. That is true all the year round, even though they only mate in one short season. This is the same social structure you find in the tiny mouse-lemurs and the very small African bush babies, so we think this is how primate social life began.'

Feeding habits and the origin of lemur families

Lepilemurs are not typical of primitive primates in one respect: they do not eat fruit. Higher up the lemur-ladder, however, there are several members of the First Family that do.

'From a lemur point of view a fruit tree is a big clump of food that ripens at about the same time,' she explained. 'And the most effective group to share your tree with is your own kith and kin, with whom you team up to drive off other lemur families.' The amount of fruit or flowers limits lemur groups to a size small

continued

Extinct Lemurs

Madagascar Island was first colonized by humans some 2000 years ago, and at least 14 lemur species became extinct as a result of predation by introduced species, particularly goats, and the destruction of the forests by Man. The most spectacular of these extinct lemurs belonged to the subfamily *Megalada-pinae*. They were animals the size of orang-utans and unfortunately the entire family has been destroyed.

Above: *Lepilemurs – keeping the secrets of the blueprint of primate social behaviour. They are thought to be the animals that most closely resemble the first lemurs to float to Madagascar on rafts of detritus, before Africa and Madagascar moved further apart.*

ALISON JOLLY
Lemurs in Madagascar

enough to allow them to feed together, but, given that, the larger the group the better, in order to fend off competition.

'We think that the First Family began in fruit trees. The primate families went on from there into the different food habitats and developed different sizes and types of families: huge groups where there were vast stretches of grass, and tiny ones in the sparse trees which could only provide food for one or two animals at a time.'

There are other reasons for family life evolving, for example, it has benefits when it comes to dealing with predators. The more eyes, the more likely the family is to spot their enemies: the Madagascan fish eagles and harrier hawks. There are also more voices to raise in a pandemonium of shouts, screams and lemur bellows and, of course, more ears to hear these alarms.

Social animals can also mob predators whereas small and solitary ones cannot. Different species of solitary lemur will gang up to see off something like the lynx-sized fossa (while single little lepilemurs must be content to hide).

The importance of females
The breeding females are the core of the family, with the males checking in and out – even that is for the good of the family. 'It's very important to prevent incest and inbreeding,' Alison observed. 'We have rules about that in human society too.'

'In all the lemur societies we know, females take precedence over the males. The females aren't more aggressive – they fight and threaten much less often than males – but if a female wants a piece of food she gets it, whatever the male wants. That is very, very unusual among monkeys and apes.' She smiled: 'I'm afraid that in most of our near relatives, males take food first, and females hang back.'

No one really knows why male chauvinism is less well developed in lemurs. 'It is only one of the mysteries. Females do need an enormous amount of energy to bring up a baby; most lemurs have just one infant a year and put a great deal of time and effort into bringing it up. A young cat, for example, can have her first kittens at a year old, whereas a young lemur of the same size has her first infant at three and spends at least five months before weaning it. It could be that there is something so special and demanding about lemur motherhood that females need priority feeding too.'

We watched a group of mother lemurs: 'Their suckling, their gestures, the mother's care, the baby's reflexes and even their weaning, are so much like ours,' Alison said almost wistfully. 'In so many ways they do seem like the First Family; the innocents in the Garden.'

Natural population control

In the 20 years that Alison Jolly has been studying the two primate groups on Madagascar (the lemurs with delight, the humans with anguish) the lemurs have maintained their delicate mutuality with their habitat, while the human population has increased from five million to nine million!

'We can trace our evolution from the prosimians that first began to specialize in learning, long life, social dependence and caring for their young,' Alison Jolly believes. 'We can see that the lemurs, relative to other mammals, have few young, buffered over a long life and varying environment.

'Why have we, the outcome of the primate line, failed to learn this lesson?' she asks in despair. 'Will we learn it soon enough to ensure our species a future?'

ALISON JOLLY
Lemurs in Madagascar

12. UNICORN-WATCH

MARK STANLEY-PRICE:
Rescuing the oryx

Nothing could be more extinct than the unicorn, indeed this magical beast with its single, spiralling horn and whimsical nature may never have existed at all outside of fable.

Or is it possible that this graceful, milk-white antelope in the photograph is that legendary creature, known to today's nature-watchers as the Arabian white oryx. Old legends of Arabia equate the oryx with the unicorn and it is quite possible that a fable of a single-horned beast could have grown up around rare glimpses of oryx in the heat hazes and mirages of the scorching deserts that are their home. Thanks to the efforts of a British husband-and-wife team, Mark and Karen Stanley-Price, and the personal generosity of the Sultan of Oman, we now have the opportunity to study the oryx and perhaps resolve the enigma of the unicorn.

The rescue of the Arabian white oryx is a success story unrivalled in the history of conservation. The programme now being supervised by the Stanley-Prices, in one of the world's most desolate places, is creating ground rules for wildlife rescues which will be invaluable in the future with other difficult animals.

At first glance their formula appears simple; in practice it is a formidable amalgam of diverse interests: that of world conservation which first recognized the dire status of wild white oryx, the organizational abilities of local conservationists in Oman, the time and dedication of field ecologists like the Stanley-Prices, the involvement of the local inhabitants, and the sustained interest of the ruler, the Sultan of Oman (not forgetting large allocations of his oil revenues).

Lose any one of these elements and the oryx would have remained as extinct in the wild as the unicorn. That was how things were in the deserts from 1972 onwards – with not a single wild oryx anywhere in Arabia.

Success story
Ten years earlier, the international conservation movement had seen the writing on the wall. Britain's Fauna and Flora Preservation Society (FFPS) gathered

Above: *Overseer of Operation Oryx, Mark Stanley-Price: he and his wife, Karen, live deep in the desert of Oman, and are trying to ensure that an antelope once extinct in the wild is re-established in its rightful home.*

Below: *Wadi Yalooni, Oman.*

Opposite: *The legend of the Unicorn lives!*

MARK STANLEY-PRICE
The oryx

the forces of the World Wilflife Fund, other conservation charities and some far-sighted Arabian rulers, to launch a rescue called Operation Oryx. A team led by Kenya's chief game warden, Ian Grimwood, caught three wild oryx on the Oman borders. To these were added zoo animals from London, Kuwait and Saudi Arabia, creating what was called the World Herd; nine Arabian white oryx that were then shipped to Phoenix Zoo in Arizona which had agreed to fund a captive breeding programme.

The last wild white oryx in Arabia were wiped out by a hunting party in the Central Desert in 1972 (the party had entered Oman illegally). In 1974, the Sultan decided to make the re-introduction of these beautiful and extraordinary animals (they can live without ever drinking water) a personal *cause célèbre*. By 1978 the World Wildlife Fund had carried out a feasibility study under the direction of the Sultan's conservation adviser, Ralph Daly, and the Trustees of the World Herd had agreed to donate animals for the re-introduction attempt.

Work was started on an enormous, one-kilometre-square fenced enclosure at Jiddat Al Harasis, just south of the Empty Quarter Desert, and in March 1980 the Sultan's skyvans flew in the first five animals.

Mark and Karen (newly-married) were nervously awaiting this arrival. Their anxiety was understandable. A complete village had been built at Wadi Yalooni, with its own generating plant, airstrip and workshops. A squad of local Harasis tribesmen had been recruited, trained and equipped with uniforms, radios, jeeps and rifles to watch the animals round the clock. Never in the history of conservation has 'so much been spent on so few'. Moreover the Stanley-Prices were well aware that the new arrivals were hardly ideal subjects for re-introduction. They had all been born in the zoo and none had any experience of the primary condition of life for a wild desert oryx – the ability to obtain sufficient moisture from dew-coated browse.

Less than a year later, however, and almost exactly 20 years after the oryx was lost to Arabia, the Harasis chief warden opened the gates to the enclosure and all concerned watched this first group of animals walk off into the bleak desert. We visited the project one year after that, a month after a second group of eleven animals had also been set free. By this time the Stanley-Prices knew that this huge gamble of time and money had been won.

Just a few hours after we had landed at Yalooni (in the same Skyvan used by the first oryx), Mark urged us out into the roasting desert (where the noon temperature had just touched 42°C in the shade) to introduce us to his pride and joy, a calf from the first herd. 'That is a totally unique little animal,' he said with undisguised pride. 'Born in the desert and never had a drink of water in his life!'

Back at Yalooni, Herd Two appeared less than delighted with their freedom and were obviously not moving far away from their familiar enclosure. It was cupboard love, Mark explained, the animals were still getting some fodder, and were learning the hard way to survive without drinking water.

'They've now been out for three weeks,' he explained. 'And we only give them water every other day. It's all part of the process of conditioning them to the realities of this harsh environment. If on the second day (a waterless day) they feel hot and dry, they learn that it is wise to go into the shade a bit earlier. We're training them for a life that is completely independent of our assistance.'

Five animals in Herd Two were born at Wadi Yalooni, and the Stanley-Prices now believe (on the evidence of the last two years) that the physiological adaptation of the white oryx species to life in deserts has not been bred out by two decades in zoos. Behavioural adaptation appeared to be an even bigger problem: would animals used to the comforts of zoo life adapt to the very harsh realities of the desert?

'These were animals brought up from birth, eating hay and grass in a small enclosure,' Mark pointed out. 'Would they actually know how to graze in the wild? We simply didn't know. In fact these animals started to graze within minutes.'

Another potential problem was that none of the imported oryx had the vaguest knowledge of their surroundings. 'Again we have been very fortunate,' Mark confessed. 'In spite of the fact that none of these animals, and that includes the dominant adults, have any idea what is over the horizon, their natural instinct seems to be to explore everything very slowly and carefully.'

The decision to move is apparently taken by the dominant female, and the rest of the herd is moved on to keep up with her by the dominant male. 'Oryx are always looking up and keeping in touch with the rest of the herd. Sometimes, if the dominant male doesn't want the herd to move he will cut round in front of

MARK STANLEY-PRICE
The oryx

Above: *Born to the desert, this young oryx has never drunk water in its life. Oryx obtain their water from the moisture in desert vegetation which is regularly dowsed with heavy dew. Nonetheless, oryx need to be knowledgeable about when and where to feed to thrive in such harsh conditions. Surprisingly, even zoo-bred animals seem to know instinctively what to do.*

MARK STANLEY-PRICE
The oryx

Below: *Saving the Oman oryx is also saving local desert communities. Harasis tribesmen have been employed to mount a 24-hour watch over the oryx and, complete with modern equipment, uniforms, and deeply vested interest are a proud force to be reckoned with.*

the leading female and herd them all back.

'The males have a hard time when they first leave the enclosure,' Mark acknowledged. 'They've been used to their females being confined by the fence in an area of about a square kilometre. Outside, they are faced with the prospect of their herd disappearing anywhere, so they get very dominant and aggressive. The dominant male works like mad to stop them exploring too far, which has helped us.

'We have had a lot of luck,' he admitted. 'When we released the first herd there had been no rain for six years and the vegetation was almost non-existent. We were feeding very heavily and we thought this might have to go on for years. In fact it rained torrentially three weeks after the release, up came the green grass and after that they were independent. We now know that we were probably worrying unnecessarily: these animals quite obviously know how to detect new areas of grazing after it has rained. They will make large-scale movements into unknown territory in order to come out on the other side where the grazing is good. They wait for the wind to change, bringing information about soil moisture, perhaps, or the smell of new growth.'

Water and grazing are the parameters dictating the way a white oryx plans its day, as the Stanley-Prices discovered by watching the first release round the clock. The Harasis guard camped out within sight of Herd One and reported their every action by radio. The same goes on today with Herd Two. All this information is carefully logged and the Stanley-Prices are now well advanced with the world's first comprehensive study of this rare creature.

'The first thing the herd had to do,' Mark said, 'was to calculate the benefits of walking back to Yalooni to get a drink, and this was entirely dependent upon the temperature. On very hot periods they would walk in every two days, drink, then move straight back out to the grazing ground. As the temperature cooled the interval spread to three, four and five days. Then at the end of the first summer they detected an area of good grazing 25 kilometres from Yalooni. The temperatures were dropping and the grass was green but they still walked back to Yalooni for a drink, taking two and a half days on the round trip. But they never did it again! By staying in the shade a bit longer, and feeding more at night, they worked out they could get more water with less effort than by trekking back to Yalooni. And from that time on they have been independent of drinking water from us, although they do get to drink physically when it rains.'

We went to see Herd One in April, 1984. 'They haven't had a true drink since last August,' Mark observed with understandable pride.

Herd Two, following in these first footsteps, will probably do as well, if not better, than Herd One because everyone at Yalooni is much more experienced now. The project has entered a new phase, with huge amounts of information accumulating about desert antelope in marginal habitats, that no other naturewatchers have had the time and money to do so comprehensively. The project is also learning how to perfect a ranger organization from scratch, using local people who are increasingly aware of their vested interest in the white oryx – a wild, free animal.

Perhaps the single most important feature of Operation Oryx is the way so many different elements have subscribed to its success; conservationists, fundraisers, field workers, bureaucrats, an Arabian prince and the most needy of his people. God (or Allah) help any foreign hunting party that tries to slaughter these oryx!

13. GORILLA-WATCH

The apes became established about 20 million years ago in the era known as the Miocene; at that time drier conditions developed and the huge grasslands sprang up. The seas were warm, the rains regular and a huge equatorial forest stretched from the west coast of Africa to the edge of Asia. Tiny mammals the size of rodents inherited this gentle habitat from the reptiles, and these mammals in the trees evolved into the primitive primates, the ancestors of humans.

Just as no single theory in natural history has aroused more interest and controversy than Charles Darwin's suggestion, in *The Descent of Man* (1871), that we are descended from the apes, so no contemporary branch of nature-watching attracts as much interest from the public as ape-watching. In the last 25 years, nature-watchers have turned to the living ape species to see what they can learn about human behaviour, by observing that of our evolutionary relatives (also with a feeling of alarm that there are so few left). Two primates have been given particular attention: the very human-like chimpanzees, and the gorillas.

In 1984 we climbed up a trail, already well-trodden by nature-watchers, in search of the Rwandan mountain gorilla. Rwanda lies deep in the centre of Africa. Its population is as dense as that in the English suburbs, and although the vegetation is strange and exotic, the higher you climb the closer you get to an English climate.

But it was an adventure! Less than 100 years ago no European had ever set foot in these forests on the slopes of volcanoes, which were the hunting grounds of cannibalistic tribes.

The gorillas were discovered by a German military officer, Captain Oscar von Beringer, on a climbing expedition amid the Rwandan volcanoes in 1902. He proudly shot two of them. Captain Oscar's trophies were later identified as an unknown subspecies of gorilla (there is a more numerous Lowland Gorilla (*Gorilla gorilla gorilla*) in other parts of Africa) and they were named after him: *Gorilla gorilla beringei*.

Very little was known about these gorillas (which soon became known as mountain gorillas) until the eminent nature-watcher, George Schaller, conducted

---

THE GORILLA

Of the family *Pongidae* (great ape) the gorilla is the sole member of the genus. There are three races, the Western lowland gorilla (*Gorilla gorilla gorilla*) living in the forest (some at sea level) in the Cameroons, Central African Republic, Gabon, Congo and Equatorial Africa; the Eastern lowland gorilla (*G. g. graueri*) in eastern Zaire; and the mountain gorilla (*G. g. beringei*) in Zaire, Rwanda and Uganda where they inhabit montane forests up to an altitude of 12,500 ft.

The world population of gorillas has been estimated at 13,000; 9000 in West and Central Africa, and 4000 in East Africa. Less than 400 of these are mountain gorillas.

Gorillas live about 35 years in the wild in close social communities which (in the east) normally comprise an adult dominant male (the silverback), some three adult females, and three or four offspring of different ages. In the west, gorilla groups rarely exceed ten members but in the east, groups of up to 30 have been observed. Gorillas are the most socially stable of all the apes. They are also regarded as being, together with the two species of chimpanzees, the animal most closely related to Man; indeed they are regarded as being nearer in an evolutionary sense to Man than the animal most closely related to Man; indeed they are regarded as being near- and the most sexually dimorphic: silverback males are almost 6 ft tall and weigh as much as 400 lb, while the females are on average a foot shorter and 200 lb lighter.

Opposite: Gorilla gorilla beringei — *shadowy giant of the steaming forest.*

ROGER WILSON
Mountain gorillas

Above: *Jungle ecologist, Roger Wilson; working with the mountain gorillas in the forests of Rwanda in central Africa, where the legend of King Kong dies hard.*

Below: *Rwanda, Africa.*

a detailed study in the 1950s. Unfortunately, for the whole of that intervening half-century, the mountain gorillas had inherited the ferocious reputation of their lowland cousins, itself grossly exaggerated, and were treated accordingly.

What we were to see was one of the great success stories of conservation – a point we shall come back to later. Deep in his very green jungle a shy, very social creature lives in the kind of family empathy that would be the envy of most 'civilised' human groups. Thanks to the very detailed work of nature-watchers the extreme vulnerability of the gorillas has been recognized and a complex, well organized strategy implemented for their protection. This works so well that the Rwandan mountain gorilla groups have remained relatively stable for almost a decade, and indeed have become so used to nature-watchers they now tolerate visits by tourists who happily pay substantially for the privilege.

The *Parc Nationale des Volcans*, where the gorillas live their gentle lives, was established in 1925 and is one of the very few game reserves in Africa which operates at a profit. If you can make conservation profitable without corrupting the wildlife, the battle is very nearly won! The Rwandan Mountain Gorilla Programme is also a rare and remarkable example of the international conservation movement cooperating successfully; the work and the costs are shared by three foreign groups: the International World Wildlife Fund, the American-African Wildlife Foundation (who contribute most) and the British Flora and Fauna Preservation Society (FFPS).

It was FFPS's young field worker, Roger Wilson, who acted as guide on our visit to Rwanda.

ROGER WILSON:
Gentle gorillas

Roger is a brusque young Yorkshireman who makes it quite clear to his tourists that he regards them as a necessary evil.

Before starting up the mountain, visitors are given a lecture by Roger about keeping still when in sight of gorillas, not touching them, not waving to attract their attention, not smoking, not leaving food about and, finally, what to do in the event of a charge. This is all taken in with much good will and not a little vicarious anticipation.

'If one of the young ones comes up to you,' he says

'don't encourage it. We don't care what you catch from them but we're very worried about what they might catch from you.

'If the silverback (the group's dominant male) does charge, go into the submission pose. Get down and look away. It's highly unlikely that he will complete the charge but you might get thumped, and being thumped by a 400 lb silverback is no joke.'

Gorillas spend their lives in what amounts to an eternal late English summer, lapsing into autumn chillness in their cold season. Four groups have been habituated to accept tourists, a number of others tolerate the regular presence of a research worker or two, and there are several wild groups. Roger Wilson has worked with all the groups and was responsible for the habituation of the most recent tourist group, the Susa. We devoted our time to Group 13 which has been accepting tourists for some time.

Apart from habituation, nothing has been done to change the gorillas' behaviour; in fact the programme goes to great pains to ensure that this should not happen. All the groups, including the habituated ones, roam freely through the park, and every morning Roger and his guides have to find them. This involves climbing to the point where the gorillas have been the day before and then tracking them (which can take up to eight hours) to their new feeding grounds. They are not difficult to track because gorillas are messy eaters, but much of the climb is on hands and knees through bamboo thickets and tangles of aromatic plants, including some painful ones such as the giant stinging nettles, as tall as a man.

Our gorilla expedition

The first encounter

We took nearly four hours on our first day to reach the feeding ground of Group 13, working our way up and along the side of two volcanic craters, behind a guide who cleared the way with his machete. Towards mid-morning when we were feeling slightly sick from the altitude (9750 feet above sea level), Roger stopped us while the guide went forward, grunting gently to warn any gorillas of our approach.

There were none about, but we had found their nest of the night before. It was on high ground and the vegetation had been trampled into a springy mattress. Roger announced that we were definitely

ROGER WILSON
Mountain gorillas

on the track of Group 13 after he had examined a large pile of droppings, which he appeared to recognize. When questioned about these he nodded matter-of-factly and said it belonged to 'Mtoto'. 'It's the size that would be made by a late juvenile or subadult, and because there's only one animal of that size in Group 13, I know those are Mtoto's.'

This may sound a bit crude, but the trick has been used for years by researchers to make sure they are on the trail of the right group; this can be important because sometimes trails cross and you end up with a wild group.

'Once or twice a year the trackers fail to find the

Above: *Small groups of tourists, heavily supervised, can now visit wild gorillas habituated to the experience. This ensures the forests are regularly patrolled, and provides work and valuable income for locals.*

gorillas, when the trails have crossed with a wild group which have gone too high. But if you find the nest and the droppings you generally know who you are following,' Roger explained.

It occurred to us that the trails could cross *after* you had found a nest, but we were too out of breath to worry about such a detail. Roger noticed our condition and allowed us to take a rest in a beautiful glade that could have been in the Lake District.

Looking around, he announced that we were sitting in a gorilla feeding site and that the animals were probably no more than 250 yards away. 'There were about ten gorillas here. They've been feeding on the giant thistles, eating the roots. That's wild celery – they pick off the skin and eat the inside. Here's cow parsley, they eat a lot of that, and goose grass and forget-me-nots.'

We found Group 13 an hour and a half later: they were clinging to the sheer side of a black cliff some 40 feet above us, munching the ubiquitous wild celery. We hauled ourselves up by these roots to within ten feet of the group and waited. Within minutes three black balls crashed down and unrolled themselves at our feet, bright yellow eyes turned in our direction, and it began to rain. It was hard to escape the impression that the gorillas, in their element, had made these arrangements in order to watch us! We sat in our ponchos for the allowed hour (another of the ground rules). It says something for the experience that none of us can remember being anything other than enthralled.

The first sensation you feel is that it cannot really be happening. The gorillas are all around you, behaving so naturally it is as if you have become invisible. They approach to within a yard or so, sometimes less in the case of the very young animals; they eat, sleep, play, groom, move amongst you, bound over you, then in their own good time, move on.

Seated somewhere in the middle of all this is the huge silverback – an animal with a body like a barrel, hands like bunches of black bananas and formidable teeth in a massive slab of a head. Now and then he will cast a polite, almost delicate glance in your direction while breaking off branches of bamboo with treetrunk arms that could just as easily tear a human limb from limb!

The problems of habituation

In the dying light Roger told us a little about the gorilla group around us and the problem they share with other groups – habituation to humans. 'Each group has its own home range of about four square miles which overlaps with those of neighbouring groups, and in which they move around, feeding as they go.' The movement of the gorillas is unrestricted but this freedom makes them highly vulnerable to

ROGER WILSON
Mountain gorillas

poachers who slip backwards and forwards across the borders. There are now about 350 gorillas left, 250 on Roger's side (Rwanda) with another hundred or so in Zaire. Their conservation in Zaire is at best shaky and little is known about the Zaire population. One hundred is a tiny population by any standard. Unfortunately, in this context the gorillas are not ferocious enough for their own good, and were far too easy to locate even before groups were habituated on visitors.

'Their defence is such that when they're approached by someone they consider dangerous, the silverback charges in order to threaten and to hold the danger off. The rest of the group flees, and when all is clear, the silverback follows them. That is what this group would have done before it was habituated.

'Habituation is simply getting the group used to people, so they have the habit if you like. The process has been developed by several people over the years as a scientific technique for observing the gorillas. You go to the group every day. To begin with the male will charge while the group flees (the habituator having adopted a suitably submissive posture), but eventually they recognize that you are not dangerous – that you are something which comes along in the forest every day, like an antelope. Then you start to move closer and closer. This takes months.

'The idea is to habituate your silverback. Once he accepts you, all the other members of the group will accept you too. Some accept quicker than others. There is one group we tried habituating, and the silverback just won't accept it. But that's all right, he can stay wild if he wants to.'

We suggested that habituated groups, which are essentially creatures that have had their natural fear of man subverted, must surely be more vulnerable to poachers.

Roger agreed: 'It's a calculated risk. In order to protect these animals we have to know all about them, get in and count them and see that everything's all right. If the poachers know we are coming every day, they tend to stay away.'

Roger also has the problem of becoming over-habituated on the gorillas: 'It's practically impossible to remain indifferent in a situation like this,' he confessed. 'You can become very involved with the life of your gorilla group. It's such an honour to be accepted this way by a wild animal, you tend to want to get closer and closer which you must not do.'

Gorilla character

On our second day out we were more fortunate in finding Group 13 within an hour of discovering where they had nested overnight. There are few flat places on the mountainside, so Julian and Roger borrowed the silverback's old nest for their first interview. Halfway through the conversation all 400 lb of silverback suddenly dropped out of a tree, bounded politely over them and sat a couple of yards away, munching. Roger played gorilla and ignored the gentle giant who had decided to join the interview.

'So much for the ferocious gorilla,' Julian smiled with just a hint of nervousness.

'Yes, it's a pretty gentle life,' said Roger. 'They've fed all morning and now they're getting ready to take a siesta.'

'Not doing much?'

'Reflecting on life,' said Roger, and he meant it: anthropomorphism is almost impossible to avoid when the creature is one of our closest relatives. 'You see so many traits that you imagine to be human traits

Above: *The film* King Kong *was written by the inventive Edgar Wallace who did some research and came up with the right jungle setting (complete with volcanoes) for his misunderstood monster. Unfortunately for the reputation of gorillas, their gentle, retiring natures were deliberately distorted for the box office.*

ROGER WILSON
Mountain gorillas

in the gorilla's behaviour.' He nodded in the direction of the silverback. 'See how he's scratching himself, that's because he's a bit uncertain. At other times he'll scratch the side of his nose, reflecting, thinking about something; looking at a new object and trying to figure it out.'

Roger has no doubt that the gorillas possess what he calls 'a very respectable intelligence'. 'I've become more and more impressed by it. It's not immediately evident how clever they are because they don't actually do that much; they don't use tools but they are very aware. If the same people visit a group regularly, the gorillas get to recognize individuals. When someone returns later after a period away, the gorillas quite obviously recognize that individual. The same thing happens if its someone they don't like – in which case they recognize them and shy away. They know all the guides and very often you'll see the gorillas watching the guides rather than the tourists.'

The gorilla way of life
In the days that followed we seemed to become accepted too, and Roger had the opportunity to describe the structure of Group 13 and its relations with its neighbours.

A gorilla's future in the group depends on its sex. When a young female matures at seven to eight years old she will generally leave her group when another group comes into range. 'She never sets out on her own but simply slips off to join the other group, which is known as a transfer,' Roger explained. 'Young females are socially immature so they often move several times. Eventually she will settle for one group, will mate with the silverback and have offspring of her own. That tends to fix her in the group. She will then only move if she isn't getting enough attention from the male. From a species-survival point of view it makes sense to move into different groups because her genes will be spread; it's like "not putting all one's eggs in one basket".

'Young males, on the other hand, are ruled by another set of imperatives. All the young males are called blackbacks and when they reach the age of 11 or so, they may turn solitary. They spend a number of years on their own, following other groups through the forest, building the strength and experience needed to attract females and become silverbacks; the silver-grey coloured back is the badge of maturity and authority.

'The other chance a male has of procreating is to stay in the group in which he is born (the natal group), in the role of spare male. His father is still the dominant male (the silverback) but if something happens to him, then the son inherits all the females in one go. We saw this in practice with Group 11 in which the dominant male died of old age. The other male in the group was still a blackback but within a week of taking the group over, although he was still an inexperienced and rather young male, he silvered up. So there's obviously an element of dominance in silvering; it is probably hormonally controlled.'

Below: *One male dominates a gorilla group; he is called the silverback, weighs as much as 400 lb and stands almost 6 ft tall. The silvering is thought to be a sexual hormonal reaction triggered by an adult male achieving dominance over a group.*

ROGER WILSON
Mountain gorillas

Gorilla conservation and the future

As our stay in the forest continued we began to recognize the individuals of Group 13: one young male called October could not resist his reflection in our camera lens and ruined many a shot. Was it possible that young October was vain as well as curious?

'They definitely have personalities,' Roger confirmed. 'There are very tolerant gorillas, and very aggressive ones; I'm sure it's character. In some cases, of course, their temperament has a lot to do with their past history.' This comment brought us sharply down to earth. Our days amongst the gentle gorillas had

Above: *Who watches who on* Nature Watch? *Cameraman Noel Smart, eyeball to eyeball with 'October', a young male who could not resist his reflection in the lens. Keeping these inquisitive animals at a distance — to avoid them catching human illnesses — is the one drawback to the tourist initiative.*

lulled us into a state of complacency about the future. In fact, Roger Wilson and all the other field workers, the tourists, our cameras and the government guards were in this forest because these gorillas had come very close to extinction. They may be one of the most attractive animals left, but their true significance to our story is the success of their conservation.

About ten years ago it was recognized that the gorillas were facing extinction on a number of fronts.

Firstly, there was the traditional African problem of poaching, and although there is no history of gorilla meat poaching in Rwanda, snares are set for other food animals and the gorillas do get caught in them.

'Comparing the groups we monitor every day – those we can free of traps – and the groups that aren't checked that regularly, there is a 17 per cent difference in the number of young; the difference seems to be the number of young killed by these traps,' Roger explained. 'In one of the groups, which has just seven members, one has a maimed hand and two lack hands.'

Even more dangerous is the breed of poacher that goes into the forest specifically for gorillas.

'It became evident in the 1970s that these people are after young gorillas, almost certainly for zoos. No one else would pay good money for a young gorilla other than a zoo. When I first came here, I found this idea unbelievable, but who else would want a young gorilla? The situation is particularly tragic because the poaching of a young gorilla inevitably involves the death of the silverback, since he always tries to protect his group. He is the cohesive element in the group and when he is killed, if there are no other adult males, the group simply collapses. The death of a silverback is an absolute disaster – in effect the whole family is destroyed with the death of that one animal.'

So anti-poaching regimes are given an understandably high priority in the Mountain Gorilla Programme's (M.G.P.) conservation strategy. There are two types of patrols: trackers who carry arms and are paid a bounty for the snares they collect and the poachers they capture, and the daily tourist runs. Every day these visitors are paying for a trip through the forest that has the effect of making it a very unsafe place for poachers.

The programme has also tackled what is arguably the more serious problem, the destruction of the gorilla habitat.

'Man the poacher is just one enemy', Roger said. 'Man the cultivator is the other. You are sitting in the most densely populated corner of the most densely populated country in Africa. All these people need land. Already there is very intense agriculture right up to the park edge which has not been resisted in the past. In the 1960s, over 50 per cent of the park was taken over for agricultural schemes; what remains is

ROGER WILSON
Mountain gorillas

Above: *Rwandan gorillas now have their own guards although poaching has not been a serious problem for some years. Poachers, after smaller game, damage gorillas who get caught in snares; Roger Wilson believes that unscrupulous zoos are still buying young animals snatched from the wild.*

just a thin strip running along the top of the mountains. We think there are about 350 gorillas left, but all the pressures – in particular the high population growth rate – are still here.'

So the M.G.P. workers are spending as much of their time with the Rwandan people as with their beloved gorillas. 'We take schoolchildren into the park, show them films, give lectures, and take them to see the gorillas; we tell them about the importance of the water, how the park conserves the water supplies by acting like a sponge. And we also talk about the park as a source of income for the country.'

This means tourism, which has been the subject of some controversy among scientific nature-watchers. Roger Wilson has no doubts. 'Tightly controlled tourism will work. I accept that it has to be tightly controlled. I also believe that no tourism at all would be an absolute disaster.'

And as far as we could see he is right. Like the safari cars in game parks which are ignored by the animals until someone climbs out of one, the Rwandan gorillas appear to have decided that the tourists, who bring in such valuable foreign currency, are invisible.

Given that everything appears to be going so well, how does Roger Wilson rate the gorilla's future? 'I'm an optimist. I believe that now the gorillas are recognized in Rwanda as a national resource, something which supports tourism in the country, great efforts will be made to protect them. At this moment it is in the balance. If the gorilla population decreases there are people outside the park who would be very happy to move into this habitat. Then the gorillas are finished. Personally, I think the gorillas are now so important to the country that they have a good chance. But, it could still go either way.'

AIR

14. BRITISH BIRD-WATCH

Bird-watching, above all other forms of nature-watching, is obsessive. The British ornithologist, W. P. Pycraft, said about it in 1937: 'A field which can never be fully explored: an interest will be stimulated which will be as a consuming, unquenchable fire.' British bird-watchers are among the most obsessive in the world, hence the fact that we have chosen to explore this branch of nature-watching through the eyes of British birders, and one eminent American enthusiast.

In Britain a running census is kept of our birds, from observations made at 250 sites by the British Trust for Ornithology (BTO). Thus it is possible to state definitively that some 2000 great crested grebes, 3000–5000 pairs of heron and about 220,000 gannets are breeding on and around these shores, to name just a handful of the 200 or so breeding species recorded for Britain (and Ireland).

The true obsessives, called 'twitchers', are on a 24-hour watch for new visitors (the overall national total is about 470 new species), and in particular for rare breeding birds (the rarest being a pair of snowy owls which have bred in the Shetlands since the late 1960s). We try to guard our threatened and endangered birds, particularly the large raptors (like the eagles and ospreys), as precious and irreplaceable national treasures.

Surveys indicate that some five million Britons have a serious interest in ornithology, and that 90 per cent of people who have gardens put out food for birds. The Royal Society for the Protection of Birds (RSPB) has over a third of a million members. The American Audubon Society (one of the largest private conservation organizations in the world), drawing on a vastly larger potential membership, can claim only a few thousand more. There are also almost 100,000 members of the Young Ornithologists Club in Britain. One cannot escape the feeling that they have discovered something that the rest of us are missing out on. And who could impart this sensation better than Britain's best-loved bird-watcher, Bill Oddie.

Bird Gardens

If your garden is short of birds, there are a number of ways you may increase the population by improving the habitat. A perfect garden for birds would be one that has been left wild, after certain important furnishings had been added; nest sites, cover, and a sensible food and water supply.

As well as selections of nuts and grains, readily available at garden centres, birds like household scraps and soft leftovers. Boiled rice is ideal and soft fried rice even better. In winter, supplementary fats are of great help to garden birds, especially when there is snow about. One way of handling household leftovers is to make up bird 'cakes'. Save your scraps of cereals and starches, and after a roast, set these vital carbohydrates in the liquid fat left over, using a cake casing or small foil dish as a mould. Drop the knotted end of a piece of string into the liquid mixture, and you will be able to hang the cakes in safe places later.

The creation of wild cover, especially if the cover includes a good selection of food plants, makes life much easier for our birds. Ideal cover acceptable to tidy gardeners as well as birds would be spiky fruit bushes, like the gooseberry, hawthorn, elder, rowan, cotoneaster, viburnam, and holly. Try and encourage thistles, teasels, wild oat grasses and heavy seeders to grow underneath.

In a matter of a year you will be in a position to watch the beginnings and ends of major migrations from your kitchen window.

Opposite: Britain is a nation of bird-watchers, watching feathered friends, like this grey heron, very closely.

BILL ODDIE
Bird-watching

Above: *Comedian-ornithologist Bill Oddie, who learnt to love birds round urban reservoirs in the industrial Midlands of England* (below).

BILL ODDIE:
A passport to fulfilment

Childhood enthusiams

Bill Oddie grew up in Birmingham and did his first bird-watching on a concrete reservoir within sight of the city. He soon acquired that special sense of fulfilment, characteristic of bird-watchers. 'I could conceive of giving up everything else for one reason or another, except bird-watching,' he confessed. 'Initially I chose the reservoir because it was close to my home, just half an hour's ride on my bike. Then it became mine in a very intimate way. There may not have been that many birds but I knew every one of them. My records were very comprehensive. I had no rivals.'

He kept such diligent records that he won the Bowater Natural History Prize with his workbook. Today he remembers it as 'very erudite and incredibly tedious'. However it had taught him the basic disciplines of good bird-watching: the art of making very detailed observations of small numbers of quite ordinary birds.

For a period of almost ten years, Bill Oddie made regular weekly visits to the lonely Bartley reservoir, some 1040 visits in all! What on earth did he discover there to stimulate such a commitment? In the school holidays, and especially in the autumn, he went there every day, arriving at dawn and departing at dusk.

The obsession grows

By now Bill had taken the next major step in his bird-watching career, one he heartily recommends to anyone with a serious interest. He joined his local enthusiasts club, the West Midland Bird Club. The club's annual reports and monthly bulletins allowed him to share his enthusiasm and get credit from respected peers for the hours he was putting in. Interesting sightings were recorded and spotters were credited by having their initials printed in the club's annual report. Within a year of joining the club he had the satisfaction of seeing his initials come up 50 or 60 times. These entries, some for quite minor sightings, also encouraged Bill Oddie to serve what he believes to be the right apprenticeship for a serious bird-watcher.

'A lot of youngsters today only hear about the rare bird chaser and learn about those birds rather than the common ones. I get a great thrill from seeing a

BILL ODDIE
Bird-watching

relatively common bird in an unusual context because I know about its normal habits; perhaps a willow warbler that should not be around at a certain time of year, or seeing snow buntings fly over the reservoir with skylarks. That's very rare indeed, although the birds as such are quite common.'

On another occasion he recalled: 'I looked up after dozing off to see what I at first thought were midges buzzing over my head, but they turned out to be birds at a great height. After about half an hour they came spiralling down and fanned out over the reservoir. They were black terns – a totally incredible sight at a concrete reservoir. It happened again later in the day so that by the end of it I'd seen 40 or 50 of them. That's a Midlands record.'

Soon this teenage enthusiast began to acquire a considerable reputation at the West Midland Bird Club, in fact there were those who began to think that the number of reports with the initials 'W.O.' were all too good to be true.

'A friend and I went on his scooter to a reservoir in Scotland and we saw a ferruginous duck, which is an official British rarity, with only 40 or 50 sightings for the whole of Britain. We duly reported it but for days nobody bothered to go near it, presuming, I think, that I was just an over-enthusiastic youngster. Then curiosity got the better of them and some went to have a look. It was indeed a ferruginous duck and my credibility improved considerably.'

His horizons expanded suddenly at 17 when he passed his driving test and was able to borrow the family car. Bill then set out to make block surveys of all the worthwhile reservoirs in the Midlands. During most of his holidays he managed to check the birds on five reservoirs, travelling about 150 miles a day in the process. On very good days he would visit eight different sites.

'I liked the idea of getting an impression of the birds to be found right across the middle of England, even though I was literally racing against time. I kept my initials in the record book by counting gull roosts, an amazingly tedious business. But it was very satisfying at the end of the year to see your name alongside a note like "1000 black-headed gulls roosting at Blithfield Reservoir".'

The adult habit
Eventually Bill moved on to more scientific studies of birds by learning the techniques of ringing and mist-

BILL ODDIE
Bird-watching

netting. He was taught, as everyone should be, by a qualified ringer, until he was skilled enough to obtain the necessary licence.

Birds are ringed most commonly for studies of their migration habits. The fine mist-nets allow them to be handled, which is often the only certain way of confirming the species and sex of a particular bird. Every bird has what is known as its 'wing-formula', a mark of identification as distinctive as human fingerprints.

Sightings of rare members of common bird families will only be accepted for the official record if they have been examined and their wing-formula checked. It is sometimes possible, under ideal conditions, to identify birds in the field by their wing-formula, using modern optical equipment.

Nowadays, Bill Oddie is more relaxed about his bird-watching. The obsessional days are over and he enjoys a reputation as one of Britain's leading and most respected bird-watchers. He is a member of several bird protection groups including the council of the RSPB, and he even manages to combine his hobby with his work. A number of *The Goodies* television programmes have been filmed on locations that allowed Bill to sneak off for a bit of bird-watching. One of these truancies sparked off one of the biggest migrations of 'twitchers' (the name given to people obsessed with seeing rare birds) in British bird-watching history.

'We were filming near Portland Bill in Dorset. I came back from a break, having found an American bird (a yellow-billed cuckoo). This was another record of a sort, because yellow-billed cuckoos are usually found dead after the Atlantic crossing. About 5000 people came to see that bird over the following five days, and I must admit it was very satisfying driving to work past a field with 300 to 400 people in it, all looking at my cuckoo.'

BOBBY TULLOCH:
The Shetland Islands

Bill Oddie is optimistic about what the conservation movement can achieve. In the British Isles the best example we know of 'ecological evolution' is in Shetland, the outer perimeter of the United Kingdom, closer to the arctic ice pack than to the national capital. Things used to be quite different but now during the summer whole islands are covered with sea birds. Guillemots crowd the inaccessible cliffs in tightly-packed rows. Auks and shags fill every other available niche and wherever there are grassy slopes in which a hole can be dug, puffins are found. On the island of Hermaness alone the puffin population is estimated at upwards of 100,000 birds.

An exercise in aggressive conservation earlier this century caused the last pair of great skuas (known locally as bonxies) to be guarded day and night for a number of years on the island of Hermaness. Now the island boasts several thousand birds. That is 95 per cent of the great skua population of the British Isles, and virtually the entire population of the northern hemisphere. It is a similar story with the beautiful gannet. Up to 1910 there were no breeding pairs at all. Now Hermaness has about 5500 pairs.

A bird that most ornithologists regard as the most superb of gliders, the milk-white fulmar, had no breeding record on Shetland prior to 1876, and is now their most numerous breeding bird.

The Shetland bird list and the developing Shetland economy show that material human advancement can be achieved in essentially primitive areas without detriment to the wildlife. This became evident in Shetland at least ten years ago. More recently they have survived the temptations and pitfalls that the invading North Sea oil industry drags in its wake.

BOBBY TULLOCH
The Shetland Islands

Opposite: *Our knowledge of migration prowess and much other bird behaviour has been greatly assisted by marking birds, like this curlew, with rings. You must have a licence to do this.*

Above: *Bobby Tulloch, who watches over the unique bird populations of the remote Shetland Islands for the Royal Society for the Protection of Birds.*

Below: *The Shetland Islands.*

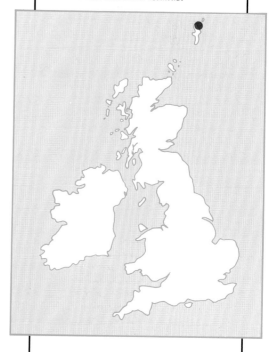

BOBBY TULLOCH
The Shetland Islands

BIRDS (*Aves*)
The taxonomic class of birds, called *Aves*, divides into 2 subclasses:

1. *Neornithes* – including all known living and fossil birds.
2. *Archaeornithes* – the sole representative is *Archaeopteryx*, the 'missing link' of the bird line. The first fossil of this strange creature was found in Bavaria in 1861. It had a lizard-like tail and many other features, including a claw on each wingtip which were more reptilian than avian. It was, however, feathered and it did fly, although not for very great distances.

Feathers are believed to have evolved from reptilian scales (which birds still have on their legs) about 200 million years ago, most likely as an improved method of controlling temperature. Modern birds have an enormous number of feathers which are mostly still used for that purpose: only about 50 of the 25,000 feathers on a swan are used for flying.

The degree to which different species of birds have adapted to different living conditions is also quite remarkable, as can be seen from a simple comparison of the habitats, shapes and life-styles of say, ostriches, penguins, eagles and woodpeckers. The act of brooding eggs, common to all birds, occurs in a diversity of ways: in nests (built almost anywhere that will hold a nest) of every shape and size; in crevices or holes in sand and rock; underfoot, as in the case of the gannets and boobies, and on the top of the foot in the case of all the penguins.

A question of balance

The birds of these savagely beautiful isles have been guarded by a native, Bobby Tulloch. A 'birder' from birth, his attitude to natural history is quite markedly different from that of other bird-watchers. Tulloch's family have been Shetlanders for generations, which means that Bobby knows almost all there is to know about the nature there.

Tulloch was the RSPB's area officer for the whole of the Shetland group (until his recent retirement); a job commanding a salary so slight that Bobby would probably have starved, were he not so adept a fisherman. His entire existence, and that of one of the world's most stupendous bird populations, is thus based on his having struck a practical deal with nature. But that is nothing new. Shetlanders have always depended upon the wild fish and fowl. Bobby is doing no more than carrying on a family tradition.

His grandfather was born on a sliver of rock called Hascosay. Bobby grew up on a larger island called Yell and enjoyed marginally more security, but life was never easy. Nevertheless the wildlife was always a source of wonder and it inspired reverence in the Shetlanders. 'In other parts it is politics or football that causes tempers to flare. Up here it's rare bird sightings, and you will do well to assume that the body making the claim is more than likely right,' Bobby told us. However, that same 'body' may well have had a puffin for his tea. Bobby Tulloch regards this as a realistic deal between man and the environment. 'In the past people had to live off the land, and that meant you had to exploit your wildlife. The price of an otter's skin was perhaps a week's wages to a crofter.'

The difference between this and what appears to be happening in much of the Third World, is that Shetlanders seem to possess what Bobby Tulloch calls 'balance'; an awareness of their dependence on the wildlife which places practical limits on attrition.

Admittedly (unlike the coastal fishermen of Africa and Asia), they also have unemployment benefit and the National Health Service; but these are recent comforts. They did not, for example, exist at the turn of the century when the great skua was reduced to the point of extinction, not by hungry islanders but by those learned Victorian naturalists in pursuit of eggs and skins.

Tulloch believes that this vital balance can be disturbed by outside interference. As a poignant example he quotes the controversial Shetland seal cull.

'It seems that all the big seal culls are done for the wrong reasons. They said the grey seals were destroying their own habitat, but it seems to me that if the

seals were left alone they would sort things out for themselves. I am not sentimental about it, but I find it very hard to justify culling to reduce the population, just because it might be affecting our fishing.'

Bobby insists that: 'We all share this planet. It has to be in our own interest to look after the other things sharing this very harsh environment.' As an example for the Third World, the Shetland story has a number of attractions. Rather than being the bleak, northern buttress of the UK, Shetland is now regarded as a unique bird haven. This reputation boosts national pride, provides considerable ecological protection and underwrites a small but steady tourist trade.

Above: *Gulls abound where humans fear to tread. Shetland's broken coastline provides an ideal habitat for birds like guillemots, who like to live between sea and sky, laying their eggs in places where only intrepid naturalists venture. Active conservation in Shetland for most of this century has produced some remarkable seabird success stories: gannets, for example have increased from 1 to 5000 breeding pairs.*

15. BIRD MIGRATION-WATCH

The little knowledge we have of how birds find their way round the world is very new to us. Two hundred years ago it was still believed that swallows vanished during the winter because they hibernated in the bottom of ponds! Even now our detailed understanding of bird migration is far from complete.

We tend, for example, to categorize birds as either migrators or residents (like the obviously migratory swallows, and the apparently resident garden robins), whereas more accurately, all birds should be regarded as migratory to a degree.

It is thus possible for every nature-watcher to study the fascinating business of migration if they are content to observe how garden birds like the robin move their aggressively-defended territories if pushed by the basic pressures dictating all migrations: the availability of food and the supply of water.

In fact the variety of movement that goes on among birds has caused activities once termed 'migration' to be broken down into a number of categories by contemporary researchers. Nowadays, these movements may be referred to as 'true migration', 'dispersal', 'nomadism', 'vagrancy', and 'eruptions'. 'True' migration is a clear seasonal shift of a population between traditional localities, there and back; 'dispersal' is a more random movement, especially of distance; 'nomadism' is continuous movement with a degree of randomness along broadly defined routes that eventually bring the bird home; 'eruptive' movements are mass emigrations to areas not normally reached; and 'vagrancy' is pure wandering into areas not normally visited by a species, usually as a result of bad weather.

We will concern ourselves in this chapter with true migrations, and the remarkable work of the British nature-watcher, Dr Robin Baker who, as a result of a lifetime of study of the mechanisms that enable animals to find their way home across incredible distances, has come to the conclusion that humans also have this 'sixth sense' (or set of senses).

Migration-watching

One of the best places to watch birds migrating is from a lighthouse, and most of the early attempts to count migration swarms were done from such coastal lookouts. When the weather is stormy, the light attracts night migrants (like larks, various waterfowl, thrushes, garden robins). Extreme conditions can produce mass tragedies as was witnessed by the keeper of the Eddystone Lighthouse in southwest England on 12 October 1901, when hundreds of birds beat themselves to death against the lantern.

A more comfortable way to witness this grand spectacle is at an estuary adjoining a narrow sea crossing. The most famous European 'bird-bridges' are the Straits of Gibraltar and the Bosphorus.

Radar detection systems have helped researchers count migrations and assess flight patterns, although they are not sensitive enough to identify individual species, especially when thousands of different birds are in the air. In fact it is often better to watch migrations at night, using a method invented by the American naturalist, George Lowery. Choose a clear moonlit night and use the disc of the moon as a background for your binoculars. You will be able to count the numbers traversing the disc and obtain some indication of the direction in which they are moving. If you do a moon migration watch as a group, each individual should take on one job; such as one counting, one recording, and another orientating. With considerable practice you can begin to recognize the silhouettes of individual species.

Opposite: *The mystery of migration once focused on swallows like this nest-builder: did they hibernate underwater through the winter?*

ROBIN BAKER
Bird migration

Above: *Robin Baker went in search of the methods and mechanisms of migration with the help of his students from Manchester University (*below*).*

Migratory senses have fascinated the human race since we first began to study animals. Aristotle (384–322 BC), who puzzled over the mystery of the absence of starlings in winter and concluded that they moved to avoid the cold, is regarded as the pioneer in the field. In his *Historia Animalium*, he plotted the movement of pelicans to their breeding grounds and speculated generally, but not that inaccurately, that cranes 'come from the ends of the world'. Aristotle must be regarded, however, as a prophet of natural history with theories well in advance of his time. Several hundred years were to pass before major advances were made to resolve the mysteries of migration. The fallacy of swallows wintering in ponds was actually pictured in a book published in 1555 (*Historia de Gentibus Septentrionalibus*) and continued to be believed by otherwise enlightened naturalists like Linnaeus, Baron Cuvier, and John Reinhold Foster who, in 1735, actually claimed to have seen swallows coming out of a river in winter. Britain's pioneer naturalist, Gilbert White, was still a convert to the hibernation theory as recently as 1767, although it must be said he had earlier come to a theory of his own that swallows flew south in winter.

In the nineteenth century, with naturalists more widely dispersed across the world, legends about the movements of birds were replaced by factual observation, and the wonder that birds actually did migrate over great distances was replaced by the even greater wonder of how they were managing to find their way.

By the twentieth century the search for explanations was proceeding in earnest. It extended outside the avian field to migratory fish, particularly salmon; other marine creatures such as turtles who had been spotted migrating across oceans, and even frogs making short, regular journeys to ponds to lay their eggs. The focus of interest, however, remained on the incredible journeys of birds who were shown by observation and the introduction of the practice of ringing (identification bands on the legs), to be travelling many thousands of miles.

More questions than answers
Dr Baker is one of Europe's leading migration-watchers and the author of several books on the subject, but points out that even today a great deal of our knowledge is speculative.

Because most birds prefer to travel at night (and feed by day) no one can yet claim to have counted migration swarms accurately. Part of the difficulty is the speed at which quite tiny birds are travelling (an average of 20 mph) in huge flocks, one obscuring the other. Many birds fly completely out of sight at a height of two miles; some geese fly as high as five miles. Nor do we know the exact proportion of the 8600 identified species of birds that are migrants. The best that can be said is that the numbers are huge. British ornithologists conducted a 24-hour count of the migration over London between 24 September and 13 November in 1960, and came to the conclusion that more than four million birds flew over the capital in that short time!

The following general guide to the migration of birds is the best we have come up with so far:

1. Most birds that breed in the high northern latitudes migrate.

2. About 30–50 per cent of birds from more southerly temperature latitudes indulge in some degree of seasonal migration.

3. Some species do both, so the key to migration must be fundamental rather than a habit of a particular species.

Marathon flyers

All migrant birds do so out of necessity: it is quite literally a matter of life or death. Admittedly some marathon flyers seem to fly further than they need to find food and warmth. They are assumed to be following dictates laid down when longer journeys were necessary, and it is theorized that the habit of long-distance migrations may have been produced by ever-longer, and more fruitful, journeys as ice ages receded or (as with some of the European insect eaters), when it was necessary to move away from ice ages.

This 'life or death' imperative is demonstrated by the fact that the hazards accompanying migrations are considerable. For Old World birds travelling south to Africa and Asia, the Sahara is never narrower than a thousand miles, and this is just the first of the several deserts they will be required to cross. There are four major mountain chains; the Pyrenees, the Alps, the Caucasus, and the Himalayas. This obstacle course also has four major water jumps; the Mediterranean, the Baltic, the Black Sea and the Caspian Sea. No bird in its right mind would attempt these journeys were it not obliged to.

ROBIN BAKER
Bird migration

ROBIN BAKER
Bird migration

What makes these journeys even more prodigious is that the bulk of the birds who migrate would not normally be regarded as great flyers. It is not difficult to image the giant albatrosses, sleek gulls and powerful raptors hitching lifts on the soaring thermals to other lands, but how about tiny birds only a few inches long?

It is hard to imagine how many wing beats the half-ounce arctic warbler makes on its twice-yearly journey of several thousand miles from the frozen north of Europe (the bird breeds nowhere south of Scandinavia) to a homeland in the heart of Asia. Its

Above: *The migratory journeys of birds can be prodigious. With numerous pauses for a drink and a bath, these golden plover travel more than 12,000 miles between North and South America every year.*

close relative, the willow warbler, weighs less than half an ounce, yet migrates 5000 miles.

In the New World, golden plovers do a round trip from North to South America of 12,500 miles; common swifts who are certainly swift but hardly substantial, are the last of the British migratory birds to arrive and the first to depart; on their journeys to Africa they feed, copulate, even gather their nesting material on the wing. Indeed, many ornithologists believe that swifts are evolving away from ever landing, if the insubstantial structure of their legs is anything to go by.

New World birds appear, according to the leading American bird-watcher, Fred C. Lincoln, to follow four distinct migratory routes (called 'flyways'): the Atlantic, Mississippi, Central and Pacific. However, some of these birds swap from the Old to the New World in their migrations, jumping the great watery ditch of the Atlantic in the process.

The feat of the blackpoll warbler

Among the most remarkable of the New World migrators is the tiny blackpoll warbler, which winters in Central America and breeds in the Canadian tundra forests. In the course of this autumn journey, the blackpoll (another bird weighing less than half an ounce) must cross the Gulf of Mexico. To do this it goes into training.

American bird-watchers using mist-nets have shown that the blackpolls take a fortnight in Massachusetts to prepare for the water jump. They feed until they have nearly doubled their weight, then on a clear day, they launch themselves off in their thou-

Becoming a Bird-watcher

If you or your children are just beginning to take an interest in nature, let bird-watching be a focus for your outdoor leisure. You have no need to be at all concerned about the quality of your bird-watching gear. Army surplus boots or wellies will suffice; adequate binoculars can be bought for about £30 and very good ones for less than £100. New anoraks are frowned upon – the apparel of novice 'dudes'.

As for reference and field guides, we suggest you start with a cheap and cheerful field guide and wait for friends to give you the glossy reference books as presents when the word gets around that you have become a birder.

A notebook is indispensable. Make notes on the spot of all your observations, keeping details of place, time, season and weather. Make a sketch of the bird, and note the size, markings, beak shape and plumage of the bird. Your notebook will also provide material for discussion and research after your day out. It is often very difficult to be sure of the species you see in the field, and if the bird is rare, only by making good notes will you have a chance to confirm a new species for your personal list.

Avoid turning your bird-watching into a form of collecting. Actually study birds, especially the ones we term common. If you follow this rule, all birds will become fascinating and you will never be short of something to watch. One day you may even be enthusiastic enough to want to travel to the Antarctic to watch king penguins, but there really is no need. Garden robins are one of the most aggressive and interesting birds (much more active than penguins) and have kept several British ornithologists intrigued for life.

continued

Left: *Mighty migrator in miniature – the blackpoll warbler – half an ounce of bird that spends its summers in South America, and winters in Canada.*

ROBIN BAKER
Bird migration

continued

A word of warning
If you come across a baby bird under a hedge, harden your heart and simply leave it to its 'natural' fate. It is most likely there because it has failed one of the tests worked out for the survival of its species; strength to fly, food enough for only one chick, disease in the nest, and so on.

It is best to rigidly follow the 'no interference' rule (but see p. 181 on bird gardens). One horned owl is now a permanent cripple, and would be dead had not Dr Gary Duke taken him in as a family pet. He has set up a bird hospital in Minnesota where he repairs eagles, hawks, owls and other birds of prey that have been damaged in various collisions with human paraphernalia. This young owl had not in fact been hit by a car while scavaging on the highway, or collided with a power cable, or been shot, as with so many of the patients of the Minnesota Raptor Rehabilitation and Repair Service. It had simply been badly overfed by someone who found it in the woods. An excessively rich, unbalanced diet had given the chick a bone condition similar to rickets – it could not walk or fly any more.

sands on a flight that can last up to 100 hours! They use half of their accumulated fat on this one incredible leg of their journey, and most of them arrive safely.

Record holders
If there were a book of records devoted to the migrations of birds, all the major entries would be held by sea birds.

The most famous marine marathon is the flight of the arctic tern (known to sailors as the 'sea swallow') which breeds within sight of the North Pole. At the first sign of the Arctic winter, the terns move out, flying almost as far south as it is possible to fly into the Antarctic. Seemingly able to walk on water, they feed on the prolific fish stocks of the Southern Ocean and complete their summer within sight of the pack ice. Their journey is at least 12,500 miles long. Every year the terns make the round trip.

On their way they overtake other great marine travellers such as manx shearwaters, fulmars, petrels, skuas, and those majestic masters of the southern skies, the albatrosses. The wandering albatross, with its wingspan of ten feet, soars in a huge circle that clips the tips of Africa, Australia and South America. In this vast global circumnavigation of the southern hemisphere, albatrosses can spend years at sea before coming ashore to join a breeding colony. A laysan albatross, released away from its colony, returned some 4000 miles in ten days!

How do migrating birds navigate?
We have only recently begun to appreciate how rich and complex the migratory ability of birds is. Admittedly we have known for centuries that birds, like homing pigeons, are remarkably adept at finding their way home; but it was simply assumed that they used their sharp eyes to follow landmarks. When serious investigation of avian migration began this century, it was quickly realized that sea birds could not be using so simple a method.

It was suggested that the techniques were either learnt or acquired by practice. This seemed unlikely when it was discovered that some of the smallest long-distance travellers rarely survived for more than two years; this was insufficient time for the young to learn the geography of their part of the world. Also, certain birds such as the European cuckoo leave their young to make their first migration alone, and these naive birds manage perfectly well.

So people started to come round to the idea that bird navigation was somehow 'built-in': part of a bird's genetic 'make-up'. This notion gained support from an experiment by the British ornithologist, G. V. T. Matthews, who released shearwaters in Boston and Venice and observed their subsequent flight-paths. They flew home to their breeding burrows in Wales, averaging nearly 300 miles per day over totally unfamiliar terrain. They wasted no time; they made no mistakes. It appeared that they both knew in which direction home lay.

Above: *How do birds find their way? Manx shearwaters released to test this question in Venice and Boston flew straight home to their burrows in Wales at speeds allowing no time for diversions. The answer is still being worked on.*

The situation isn't quite as straightforward as it at first seemed. For example, A. C. Pendeck in 1957 captured 11,000 starlings in Holland, that passed through the country every year on their way from the Eastern Baltic to France and England. He later released the birds in Switzerland after having ringed the

ROBIN BAKER
Bird migration

mature birds differently from the young ones. Most of the youngsters flew a line analogous to their original flight path and ended up in the wrong place. On the other hand the majority of the adults were able to reorientate and find their way to their desired destinations. These results indicated that the young were using raw unmodified genetic 'knowledge', which because of its rigidity led them astray; the adults were able to alter this appropriately using their past experience. So birds use both genetic and acquired knowledge to guide them on their long journeys.

The importance of the sun and stars

This knowledge, of whatever kind, must use external cues to direct the birds in the appropriate directions. The German researcher, Gustav Kramer, did some experiments in the early 1950s which indicated that birds use the sun's movements to help them to navigate. He noticed that his caged starlings grew distinctly agitated as their migration time approached. They would gather in the aviary at a point closest to

Below: *Some nature-watchers believe albatrosses (here a black-browed member of the family skims the wave tops) never stop migrating; 'wandering' may be a better term for their seemingly almost ceaseless flight. Many undertake a complete circumnavigation of the southern hemisphere, clipping Africa, Australia and South America.*

the direction their migration would normally take. Kramer believed that his starlings were 'reading' the angle of the sun and responding accordingly. More-over as they were shifting their take-off position as the sun went down, he also concluded that the birds could tell the time. To test this he set up a number of mirrors round his aviary to distort his birds' sun-sights, and he moved the mirrors to simulate different times of day. His restive starlings shuffled round their cages in perfect obedience to these shifting false suns.

Another German, Franz Sauer, tackled the problem (in the 1950s) of how birds migrate at night. If some birds could read the position of the sun and navigate using some form of time clock, were there others that could read the stars? Were there birds that could do both? A number of species (such as some warblers) were known to fly by day and night. Sauer conceived the ingenious idea of releasing birds inside a planeta-rium. He used those long-distance migrators, the warblers, and by moving the planetarium's star field he was able to distort their flight path as he chose. So this showed that the warblers were indeed using timed star sights to navigate. But was that all?

The magnetic sense
From the time we discovered magnetism, there were naturalists who toyed with the idea that, somehow or other, birds might be following the magnetic paths of the Earth in the same way as a compass needle will always point in the same direction. The idea was first put forward by a nineteenth-century naturalist, Dr von Middendorff, but it was treated with great suspicion, almost until today.

In the 1960s a number of experiments into what was still being termed a 'sixth sense' were conducted on a wide variety of creatures who appeared to have an affinity, or at least a strange relationship, with magnetism. Certain mud-dwelling bacteria were noted always swimming north and it proved possible to alter this imperative by disrupting their magnetic fields with a coil magnet. Similarly, primitive planarian worms, mud snails and bees appeared to be using magnetic clues. For a while almost everything that moved was disorientated with magnets, including green turtles, fish and many different species of birds, not least the much-investigated homing pigeon.

Finally it fell to an American nature-watcher, Charles Walcott, to prove that homing pigeons had not just a magnetic sense but an actual compass con-

ROBIN BAKER
Bird migration

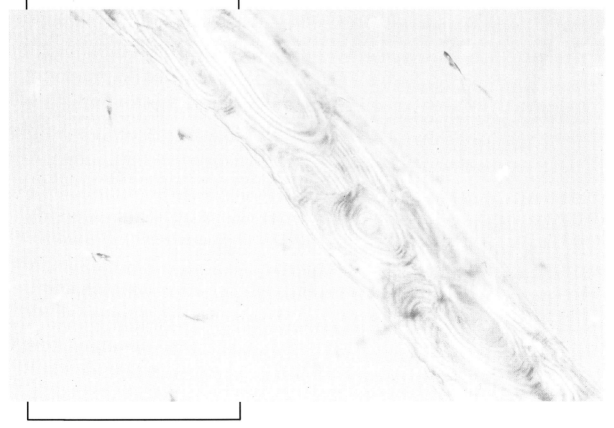

Below: *The sun and stars and even an inbuilt sense of time are all employed by birds and other creatures in their complex migrations. Recently an actual compass – polarized ferrous magnetite (the blue line below) was found in the heads of pigeons, and Robin Baker has since discovered a similar directional deposit in human skulls.*

nected to their brains. We met Charles Walcott briefly some years ago and are not surprised it was he who made this great breakthrough, for he is obviously not prepared to let much stand in his way: 'When I decided to try and find out how pigeons were navigating, it seemed obvious to me that the best way to do it was to get up and follow them. Surprisingly, no one had done that before. I managed to get some tiny radio transmitters attached to homing pigeons and we radio-tracked them from my own plane. You can imagine, that really was hairy flying! Sometimes we had to put down in strange places; on one occasion in the grounds of a mental institution. The staff found me walking around in the bushes with large earphones on my head, waving my antennae, claiming I'd just dropped in to look for my pigeons!'

In 1979 it was the innovative Walcott, working with James Gould, who found the actual compass – a crystal of material that appears to be magnetite (magnetic iron ore, Fe_3O_4, polarized to the Earth's magnetic field) inside the pigeon's head. Work has been going on ever since to establish exactly how pigeons use their micro-compasses, and there is still a long way to go.

The sixth sense

These and earlier discoveries had been intriguing a British nature-watcher, Dr Robin Baker, who, after several years of working with the magnetic senses in small animals (particularly mice) decided to tackle the very controversial question of a magnetic navigational sense in man – what had been termed 'the sixth sense' by Emile Jarval in 1905 from his work with blind people who were able to avoid obstacles.

'I was intrigued by the fact that since we started to study the magnetic sense,' Dr Baker said 'we've been finding it wherever we look. We now have a dozen or so animals (birds, wood lice, moths, snails, beetles, salamanders, salmon, tuna, honeybees and, possibly, crocodiles) who are known to possess the magnetic sense, and the way things are going it looks as if the final search will be for something that doesn't have a magnetic sense.'

It seemed inexplicable to Dr Baker that Man's much vaunted 'sixth sense' was not tied in with a natural compass, especially when he reviewed the human legends about it. One of the most remarkable of these is recorded by Captain James Cook about a native navigator, Tupaia of Raiatea, an island near Tahiti, who joined Cook on the *Endeavour* and stayed with him for a voyage of more than 7000 miles, involving a circuitous journey between latitudes 48°S and 4°N. No matter what their circumstances, Tupaia (according to Cook) was always able to point accurately towards Tahiti.

Similarly there are accounts of guides in India, and Cossacks in Siberia (Frere, 1870; Wrangel, 1840), finding their way across areas without landmarks or tracks, with what one of these observers described as 'unerring instinct'.

In the more prosaic surroundings of the north of England, Dr Baker decided to use his students from the University of Manchester to see whether he could convert this fiction to fact, employing the method that had previously been tried on birds: the deliberate distortion of the magnetic sense using magnets, in this case in special headsets.

'We looked at all the obvious possibilities first: was the sun shining and could people feel the heat on the backs of their necks, but it didn't seem to have

ROBIN BAKER
Bird migration

any effect at all on people's directional abilities. So I put magnets on their heads and, lo and behold, it disrupted their ability to know in which way they were going.'

To determine whether we use this magnetic sense on a daily basis, Baker shipped coachloads of his students to remote spots and monitored their progress towards given magnetic reference points. Not only were the majority normally capable of orientating to one of the magnetic lines, but when the poles in their headsets were reversed by magnets, a significant percentage went consistently wrong.

Julian Pettifer joined one of these outings, and had the distinction of being consistently wrong in judging which way he was facing, a not uncommon outcome of a session in the Baker 'electric chair'. However, this is not a negative result, as Dr Baker was quick to point out, for while Julian was pointing in the wrong direction (almost exactly 180° out) the fact that he was mostly wrong to the same degree indicated strongly that he was using a magnetic clue.

Robin Baker is the first to acknowledge that his finds are not all-encompassing; he has done no more than lift the lid off a treasure chest of human navigational senses. Already his tests have revealed other bizarre possibilities – like the fact that the human compass appears to be affected by the direction in which you sleep.

'We have found that if you sleep in what for you is the "wrong" direction, your magnetic sense seems to fold up. People who are consistently good are those who sleep with their feet pointing north. South is nearly as good but their magnetic sense folds up more quickly than north-sleepers; it seems not to be so robust a sense. East–west sleepers do all sorts of strange things, they can be out on a completely random basis; in fact they show no real evidence of a good magnetic sense at all.'

All of this strengthened Robin Baker's belief that the human sixth sense was indeed a magnetic compass reading: 'If, as we suspect, we are like those other animals which have been shown to have magnetically-polarized particles of magnetite in their heads, then the direction in which you sleep is a re-setting of the magnetic sense; realigning the particles of magnetite.'

As Charles Walcott had realized with his homing pigeons, Baker knew there was only one way to get this accepted by sceptical scientists (who had grouped most of this work with investigations into the paranormal) and that was to find the microcompass in the human body.

The Zoology Department at Manchester University made detailed examinations of hundreds of sections from the human skull, and finally found what they thought they were looking for in the sinuses. Just below the surface of the bone (see page 198) there was this thin layer of ferrous magnetite.

Baker concluded: 'There is no doubt in my mind that Man has a magnetic sense. We also have enough data to be clear that it is being used more or less all the time, even though we are not aware of it.'

Apart from the thrills of a major scientific discovery, Robin Baker is delighted that these findings help fit Man back into the general fabric of the animal kingdom by revealing that we still share a basic sense with birds, bees . . . and bacteria. 'If we take a backward look at the sociobiology debate from the perspective of the study of animal navigation, we can see no good reason for further ideological warfare,' Dr Baker wrote in the book which described his discoveries (*Human Navigation and the Sixth Sense*): 'Let us study Man and other animals side by side and let the academic and practical benefits that emerge speak for themselves.'

ROBIN BAKER
Bird migration

Binoculars and Telescopes
As with cameras, good magnification is important in a pair of binoculars or a telescope intended for bird-watching, and again the price you pay varies enormously. The best new lightweight German binoculars cost hundreds of pounds, while the Japanese make some that will serve most amateur bird-watchers quite adequately for less than £50. A good compromise would be a pair of high magnification (8 × 40) binoculars, preferably of the 'wide angle' type, for £50–£100, from a leading brand-name Japanese optical manufacturer. Relatively cheap second-hand binoculars are easy to come by, especially military glasses. They have the highest magnifications, but are rather heavy.

Telescopes are for more advanced enthusiasts only, though if you intend to be a heavy birder, we would recommend the Bushnell Spacemaster, with × 20 – × 45 magnification (this will cost something over £150).

16. MAURITIUS-WATCH

An ocean away from all this lies the paradise island of Mauritius. It possesses no endemic mammals nor predators (except for one bat species) because the island broke away from the African mainland before they evolved there. So for all of early history Mauritian creatures lived without fear. Four hundred years ago, before the island was settled by Man, it was an animal's Utopia. Those first sailors who walked up the beaches would have seen dodos, flightless parrots, palm savannah instead of trees and giant tortoises. Inland there were dense ebony forests: beautiful trees found nowhere else in the world, trees that took perhaps a thousand years to grow. In these forests were more parrots, of an unknown number of species. Up on the plateaux there were pink pigeons and white water hens, the latter described by an early French naturalist as standing five feet tall. We think there were also coots, darters, ducks, geese and herons, all in a myriad of form and colour. There was a red, white and blue pigeon which was meant to be good to eat (it became extinct in 1831).

In the wake of the original sailors came the settlers. African slaves, Indian indentured labourers, Chinese shop-keepers, Portuguese overseers, French farmers and, finally, British colonialists arrived on the island. They slashed and burned their way through the unique palm forests, ate species after species of birds into extinction, planted every arable inch with alien food plants (most of them proof against the indigenous insects and parasites and so were unaffected) then entrenched generations of highly evolved opportunist feeders: mina birds, rats, pigs and monkeys.

Four centuries later, with the dodo remembered only because it has given its name to the extinction threat that hangs over all of nature, that ghastly massacre is still going on. We are not levelling blame, but the fact remains that Mauritius is a tragic example of the rape of the earth in microcosm, and *cause célèbre* for bird-watchers.

CARL JONES:
Bird-watcher with a mission

Carl Jones, a young Welshman, is crusading for three Mauritian birds: the rarest bird of prey in the world,

Opposite: *A refugee from extinction: this chick and its family of Mauritius kestrels would never have made it to the end of the century without the efforts of conservationists.*

Above: *Carl Jones shows Julian Pettifer his next-door neighbour, a tenrec called Rupert.*

Below: *The island of Mauritius.*

CARL JONES
Conservation in Mauritius

Above: *A symbol of extinction that still haunts the island of Mauritius, the legendary luckless dodo (reconstructed from remains). A giant flightless pigeon that had never known fear; dodos, their young and their eggs were eaten by the first humans and their pets to hit the island's beaches.*

the small golden Mauritius Kestrel, *Falco punctatus*; the large bright Pink Pigeon, *Columba mayeri*; and the tiny green Echo Parakeet, *Psittacula eques echo*.

When luck brought us into the delightful company of Carl Jones, it soon became obvious that his interest in birds is so intense that it has the appearance of mild insanity. 'My first love was birds of prey. Not flying them or anything like that, just worshipping them. My parents tell me I've always had an innate love of them. As Konrad Lorenz (whom Carl reveres) would say: "I've become imprinted on them as they imprint on me." It's a wonderful relationship. It's been like that for as long as I can remember. Not just for birds of prey, but natural history in general, and raptors in particular. They always made my heart thump when I saw them. Over the years I've become more and more involved with them, trying to understand what makes them tick. We can sit back and study animal behaviour, we can look at it and we can quantify it, but that isn't enough. We've got to try to understand the animals too.'

He uses the word 'understand' with a peculiar secondary emphasis and a slightly shy smile. 'The only way we can really understand animals is to be involved with them and to identify with their emotions, and thus really appreciate what they are doing. I know when this falcon (his peregrine, Sweetheart) is angry. I know what its motivational states are because I've lived with falcons. When I first started keeping hawks, owls and falcons I found it difficult to understand them. But by living with them, and I mean *living* with them, you get so they are an extension of yourself. It is something you can't put down on paper, a lot of it is unconscious.'

The Mauritian crusade

So how did this Welsh birder come to be in residence under a mangrove tree in Mauritius? A tree which, incidentally, he shares with a creature called Rupert. (We mention Rupert, a Madagascan tenrec, simply to give us the opportunity to show you his picture.)

It is time to correct any impression we have given that the present Mauritian government is as hopeless as its colonial predecessors when it comes to conservation. The rape of Mauritius took place long before the present incumbents were born. The government, though financially pressed, has maintained its share of an aviary complex, designed to save the most threatened Mauritian birds. It also has a deeply committed conservationist, the Conservator of Forests, Wahab Owadally, looking after the scraps of forest that escaped the sugar barons. Owadally installed Carl Jones in his Black River Aviary in January 1979, with funds from the International Council for Bird Preservation, and the Jersey Wildlife Preservation Trust.

Wahab Owadally handed him a very simple brief. There were three Mauritian birds about to become extinct and Carl was to do what he could to save the genetic stock, using techniques not dissimilar to those employed by George Archibald in his crane-breeding programmes (see next chapter).

They both knew, however, that there was little chance of saving more than one of the threatened trio: the pink pigeon. Echo parakeets were so near extinction as makes no difference, or so it was thought, and previous attempts to save the Mauritius kestrel had failed.

CARL JONES
Conservation in Mauritius

THE DODO
Our earliest knowledge of this strange bird – 'which is *very* unlike a pigeon, but *certainly* unlike any other bird' (W. P. Pycraft 1937) – dates from Dutch records (1598), when it was described as 'Walgh Vogel' or 'Nauseous-bird'. At that time it was abundant on Mauritius.

With no endemic predators, an equitable climate and an abundance of food, the beak of this strange pigeon expanded and grew more hooked, and it ceased to fly. Eventually its wings degenerated to mere vestiges, and the 'keel' of its sternum was reduced to vanishing point. Pictures of a living dodo were drawn in the early seventeenth century by Roelandt Savory, and these indicated it had also reduced its tail quills to vestiges, leaving only the large, loose, upper and under tail coverts.

Although early Mauritians probably ate the 'tame' dodos, the Dutch description (above), which is assumed to refer to the poor taste of the meat, would appear to excuse early settlers of direct responsibility for their speedy extinction. This is now blamed on the introduction of pigs, against which dodos, their young and their eggs, had no defence.

Left: *Today's dodo? Success with other endangered birds, particularly the most threatened raptor in the world, the Mauritius kestrel, may provide support for the rescue of this Mauritian echo parakeet; at the last count there were no more than six left alive in the world.*

CARL JONES
Conservation in Mauritius

Below: *Once the scales were weighed against their species, but these kestrel chicks have almost certainly swung the balance in their favour. Eggs taken from the wild have successfully hatched and, with luck, there are now enough birds to form the basis of a successful captive breeding programme.*

We have made two expeditions to Mauritius to talk and film Carl Jones at his work. The first was in 1981, when he told us about the plight of the birds, and in particular why the kestrel was so endangered.

'First of all, if you cut down the forests you're going to lose all the birds in them; but also what has happened on Mauritius is that the very nature of the forest has changed. When the sailors first arrived on these islands, they talked about the ebony trees and the special structure of the forest – four different layers and an open canopy.

'The Mauritius kestrel is a forest-living species and it developed a special method of flight and hunting technique in this open canopy for one prey species in particular, the beautiful green day geckos that we call phelsumas. But these phelsumas are dependent upon native vegetation. When the native forest was cut down or destroyed by cyclones, the tougher alien plants which the first settlers brought in took over. The kestrels can't survive in the new forest, it doesn't provide the structure and it doesn't have enough native lizards. We're only finding kestrels now in

areas where there's native vegetation and still reasonable densities of these geckos.

'We started studying the kestrel in 1973 and we have never found more than one or two nest sites in any one year. In 1975 there was no breeding at all. So it's on the very threshold of extinction. In terms of actual numbers it is very difficult to count, because it lives in the forests and spends most of its time unobtrusively in the canopy. You could walk right past one without noticing it. But I think the absolute maximum must be 15 birds, and quite probably below that.'

When Carl gave us that assessment in the summer of 1981 he was, as he admitted later, being deliberately optimistic. His private prognosis of the status of the kestrel was that it was extinct in all but name. Too few birds were breeding in the wild to keep the species ahead of predators like the very active and abundant, introduced egg-stealing macaque monkeys.

The little echo parakeet was in an even worse state, its habitat and nest sites having been taken over almost exclusively by a much more successful exotic (imported) parrot. Carl had not seen an echo parakeet that year, and was estimating their total numbers as six. He had the bird listed as the most endangered parrot in the world (if they still existed at all). Finally there was the pink pigeon, also very rare.

Carl was in a state of some despair. He was not the first birder to attempt a rescue on Mauritius, just the youngest and most energetic. He was beginning to think the island was somehow jinxed, if not cursed, by its appalling ecological record. 'Early accounts suggest there were 30, if not more, species of bird on this island. Half to two-thirds of those have disappeared. Today there are only 11 native species of birds left. Of those 11, eight are in danger of extinction and the three we are concerned with are on the very brink of extinction.'

On the strength of that kind of indictment, it would surely have been reasonable to expect that some small emergency flare would have been fired on behalf of the sinking bird life of Mauritius, by one of the world's ecological watchdogs. On the contrary: Carl Jones was taking valuable time off to talk to us because his lovely birds had been judged a lost cause by some so-called conservationists. He had been given 12 months to tidy up and leave these birds to go the way of the dodo. The precious resources, human and financial, that Carl Jones represented would then

CARL JONES
Conservation in Mauritius

be reallocated to a cause with a greater chance of success. But he was adamant in his cause and he refused to sign the birds' death warrant.

Almost a year after our first visit, we returned to the island to see how his defiance was standing up. The Black River Aviary was bulging with an embarrassment of pink pigeons!

The Mauritius kestrel lives

Shortly before we arrived a miracle had occurred: to judge it as anything less would be an understatement.

Some two weeks previous, one of the game guards from the private hunting estate that leases most of the Black River gorges, had told Carl that he had seen a kestrel flying into a hole in the cliff, some two miles from his aviary. (Mauritius kestrels nest in rock holes, and the few remaining birds were thought to be going extinct in the total privacy of the remote volcanic crags that form the spine of the island.) Carl immediately investigated the game guard's report, and was able the confirm the sighting and to witness an incredible sequence. 'I couldn't believe the nest site could be so accessible. I came scrambling up here (a steep but passable 600 feet) and found a pair of kestrels that had just started to prospect at this nest site.'

If you are not a bird-watcher it may be difficult to appreciate the magnitude of this sighting. Carl Jones was able to sit on a rock within walking distance of his aviary and watch two kestrels working to prevent their species from becoming extinct.

When we landed in Mauritius again, the nest had been under constant observation by Carl and two young British bird-watchers who were his assistants on the island. They had set up a tent on a precarious ledge on top of the mountain and were conducting a 24-hour vigil.

Carl had already made the decision to remove any eggs that were laid. This would not be a haphazard snatch but a carefully calculated operation. 'It isn't as horrifying as it sounds, because from experience with other kestrels and kestrels in captivity, we know that if we take the first clutch of eggs the chances are very good that they will lay another clutch. It has also been noted in other parts of the world that falcons very often change nest sites. So by taking the eggs from the nest site that is vulnerable to monkey predation, the female may well choose another site which could be a lot better. We might end up with twice as many kestrels.'

We actually witnessed Carl removing three eggs from the kestrels' nest on the rock ledge. He put them in a vacuum flask packed with pre-warmed cotton wool. Then it was a race down the mountain, home by jeep and into the little room behind the aviary which serves as Carl's laboratory, and often as his bedroom.

The precious eggs were inspected with great care. Cracks can be sealed with nail varnish, but these specimens were flawless. Finally they were weighed, measured and numbered before being placed in the incubator. But that, as Carl explained, was just the beginning. 'What we have here is the fruit of literally years of work by a great number of people. I'm just the most recent to have a go. The project was started in 1973 and we still haven't established the Mauritius kestrel in captivity. The immediate priority is to look after these eggs carefully. They have to be incubated at the correct temperature and humidity. That will mean somebody staying with them virtually 24 hours a day; in fact I'll be sleeping with them at night in case the power gets cut off, then I can rush round and check that our emergency generator is working.'

He would be doing that for three weeks! In the meantime, Carl intended to patrol the Black River gorges in search of other kestrels that might have started to lay eggs. These he would remove as well in the certain knowledge that he could do a better job with the eggs than the female kestrel.

'My most optimistic forecast for the bird in the wild is that it will go extinct before the end of the century unless we intervene. Yet all it takes to raise it in captivity is a little intelligence and some good management.'

Two of the three eggs hatched. The parent kestrels whose nest he had robbed laid more eggs in a less vulnerable site. Kestrel fever gripped Mauritius for a time after the hatching and with the help of the Mauritian Field Force, Carl was able to scale another cliff from which he took another three eggs.

P.S. On 29 September 1984, the first egg was laid to a pair of captive Mauritius kestrels, and one fertile egg has also been taken from the wild pair on Montagne Zaco.

CARL JONES
Conservation in Mauritius

CARL JONES
Conservation in Mauritius

Below: *Success breeds success: Carl Jones now has a surplus of once highly endangered pink pigeons in his aviary. Success with these birds provided the incentive, publicity and funding to tackle the Mauritius kestrels and, in future, could save other crisis cases like echo parakeets.*

The pink pigeon

Carl, as we said, had been depressed about the rarity of the pink pigeon in 1981. The cause of its decline was not because it had been shot for food, indeed it was said to make poor eating. Local folklore has it that pink pigeons can be poisonous at times of the year, when they eat a certain narcotic fruit. 'They are supposed to wander around drunk, having hallucinations,' Carl grinned. 'As a result nobody shoots it, or very rarely.' Nonetheless they are desperately rare because of habitat attrition and nest predation.

On our second visit he took us to the Black River Aviary and, with extreme care, lifted a small scrap of skin, bone and prickles from beneath a cooing collared dove for our inspection. 'Beautiful,' he said in his strong Welsh accent which lends a special quality to the word. 'That's the youngest pink pigeon

in the world – isn't it beautiful? I've bred over twenty of those already this year; you might say I've got an embarrassment of pink pigeons.'

Not that it was that easy. At Black River, Carl has discovered new causes of pink pigeon rarity: they are neurotic in their courtships and make extremely lax mothers. 'Pink pigeons are the worst parents you could ever imagine in your life.' (Hence his use of collared doves as nursing mothers.) 'They drive me to absolute despair. If you want to give someone a nervous breakdown, give them some pink pigeons to look after. If you get over their matrimonial problems – which are vast – and you get them breeding, they then proceed to lay their eggs off perches, or they lay in-fertile eggs. They smash their eggs. They make a token gesture of a nest. So, if by some fluke you've got a pink pigeon that makes a nest, lays a fertile egg and manages to incubate it, you then find they don't look after it properly! The only hope is to take the eggs and put them under doves, which are beautiful model parents.

'In the wild a pink pigeon would be very lucky to rear four or six youngsters a year; that would really be pushing it. As a result they are desperately rare. I'm reasonably convinced that the wild population is somewhere between ten and twenty. That is a critically low number.' He cast an affectionate glance at his cooing doves. 'But with a little help from my friends, I can raise 40 eggs a year in captivity.'

Therein lies Carl's secret; a practical answer to extinction is production line management, to keep the curse of the dodo at bay. Carl is contemptuous of the committee-conservationists who view highly threat-ened species with pessimism.

'What we should be doing is realizing that the knowledge we're gaining with the Mauritius kestrel and the pink pigeon can, and will, be applied to the conservation of other species of birds elsewhere in the world.'

17. CRANE-WATCH

GEORGE ARCHIBALD:
A life devoted to cranes

In the 1970s nearly half of the world's 15 species of crane were in danger of extinction. This disturbing piece of news attracted the attention of two Cornell University biology graduate students. Their names: George Archibald and Ron Sauey. They were studying the behaviour, calls and language postures of American sandhill cranes.

Ron Sauey's father owned a farm in Wisconsin that had been used for breeding horses; Ron suggested to his father that this could be converted into a crane farm. His wish was granted in 1972, government recognition was obtained, and the following winter birds were installed at the farm. The first chicks were born in 1975. Today the farm provides shelter for 139 cranes and this sanctuary, the International Crane Foundation (ICF) in Baraboo, Wisconsin (USA), is now the largest captive bird-breeding programme in the world. In September 1981, we visited it to talk to George Archibald about the project.

Breeding cranes in captivity

Cranes are truly wild birds. Their flying abilities give them access to some of the most remote areas of the world. Their preference for wetlands as breeding grounds ensured privacy and protection prior to the interference of Man. But, on the other hand, very little was known about the basic needs of the crane family when George Archibald and his colleagues set up the International Crane Foundation. They learnt everything from experience, and much of it was very hard. For a start, what did cranes eat, or, more to the point, what could they be seduced into eating in captivity?

George Archibald told us: 'At first we fed them turkey pellets, but because this feed is designed to make turkeys grow fast, our cranes grew too quickly for their own good. Those reed-thin legs have very delicate joints which will bend under the weight of an obese infant bird. So we got round that problem by putting a high percentage of cellulose with no food value in the pellet. The birds fill up quickly but it slows their growth and we don't have a leg problem.'

In the course of finding out what cranes eat, the

Above: *American-Soviet crane concern overflies the Cold War in the shape of 'Aeroflot', the Siberian crane George Archibald bred from an egg taken from a marsh in the USSR to Baraboo, Wisconsin.*

Below: *Baraboo, Wisconsin.*

Opposite: *Common cranes once bred in Britain. As with many of the world's fifteen crane species, habitat disturbance and interference with their migration has caused them to move to safer breeding grounds.*

GEORGE ARCHIBALD
Cranes

Foundation was also filing away new information about the incredibly rapid growth rate of the birds. A week-old chick that could safely be contained in the palm of George's hand would, four months later, stand as tall as a man and have a wing span of nine feet. The eastern sarus crane is the tallest flying bird in the world.

The whooping crane

When we met George Archibald at the ICF he was running to and fro on a dusty path, waving his arms like a dancer in *Swan Lake* and directing soft 'whoops' at a three-feet-tall female whooping crane. Her name was Tex and she had been bred in captivity at the ICF. As she had been reared by humans she had become imprinted on them and was reluctant to mate with her own kind: so George had learnt the mating dance of the male whooping crane to get Tex in the mood for artificial insemination.

This is just one of the problems that faces the ICF team and one that they have solved successfully. Imprinting birds on humans is usually inadvisable (as we have said elsewhere), but in this case there was really no alternative. As George explained, Tex would be simply too vulnerable in the wild where, at the last count, there were only 120 whooping cranes left. 'Tex is the last of her line; the ICF has made an intensive study of the Texas-based whooping cranes and we know that Tex has no next of kin left.'

Had you been at Baraboo in 1982 you would have witnessed Tex and George installed in the family cage. In the spring George put a little desk, a cot and a telephone in Tex's house and spent six weeks with her. They brought in semen from a whooping crane breeding colony in Maryland and inseminated Tex artificially. We are very happy to report that the honeymoon paid off. Tex reared a healthy chick that autumn.

The eastern sarus crane

The ICF could not function without the help of an equally committed group of amateur bird-lovers known at Baraboo as the 'bird-mothers'. Bird-mothers play a vital role in the rearing of another highly endangered species, the eastern sarus crane.

At Baraboo, the bird mothers are on duty for an incredible 14 hours a day. For a period of three months they feed and water their young charges, weigh them, clean their pens, exercise them outdoors and break up the constant fights between the chicks.

One of the reasons these beautiful birds are so scarce is the tendency of the chicks to fight – literally – to the death. The behaviour is instinctive, a classic case of 'survival of the fittest'.

In the wild, sarus cranes normally lay two eggs, but there was rarely enough food available for every adult to successfully raise both chicks. It was left to the chicks to sort it out; this natural selection imperative has left sarus crane chicks with ferocious natures.

The ICF is of course delighted with every chick hatched, and they have called in local bird-watchers to mother the youngsters until they have fledged. Youngsters then naturally recognize that their brothers and sisters have made it through the crucial infancy period; everybody calms down and can be cared for in flocks.

The bird-mothers also conduct school and public tours of the ICF, informing visitors that there are 15 species of crane in the world and Baraboo can boast 14 of them! They point out that seven of these are in real trouble in the wild, mostly as a result of conservation malpractice in the comparatively recent past.

The Siberian crane

Unlike most other species of crane which become reasonably friendly with the humans who look after them in infancy, Siberian cranes remain vicious all their lives.

Prior to (and during) meeting us, George allowed a Siberian crane called Aeroflot to work off some of this territorial aggression on an old broom, held at arm's length. Although it looked funny, it is deadly serious. In the wild, Siberian cranes defend territory so strenuously that some birds die of wounds.

'Normally what will happen is that the intruding crane is at a psychological disadvantage, and the pair that are already utilizing the territory will defeat the other bird. Most serious fights occur at the edges of territory. If, for example, two males meet at the edge of adjoining territories, both will be very aggressive because both in a sense are "in the right".'

Aggression can also get tragically misplaced, as once happened at Baraboo. 'We had a male who got very aggressive towards another male in an adjoining cage, but, of course, he couldn't do anything about it. So what happened was that in the time-honoured tradition of the man who comes home from work after a row with the boss and beats up his wife, this crane attacked his spouse so badly that she died. We've

GEORGE ARCHIBALD
Cranes

continued

Cranes fly high and fast. Radar tests have recorded speeds of 30–55 mph, sometimes at altitudes in excess of 6000 ft. The birds can stay in the air for hours at a time. With the exception of crowned cranes (which favour trees), cranes roost at night on the ground or in shallow water, often in huge numbers (as many as 30,000 birds).

The cranes are a highly-threatened family, with two full species (whooping crane and Siberian crane) endangered, three (Japanese, hooded and white-naped) vulnerable and one, the black-necked crane, indeterminate. One race of sandhill cranes (the Mississippi) is now also regarded as endangered. Cuban sandhills, the wattled crane and the eastern sarus crane are all thought to be declining.

GEORGE ARCHIBALD
Cranes

decided we can't trust these birds, so we keep them apart and employ artificial insemination and incubation.'

These techniques have other advantages for the survival of crane species. The ICF has managed to get birds to lay more prolifically than they would in the wild. The traditional pair of eggs has been increased to as many as 18 per year by removing eggs for artificial incubation. These human-nurtured eggs have a much better chance of hatching.

Tricks are also played on the cranes to keep them laying. In the wild they breed as far north as 75°N, up in the Arctic Circle, where there is almost eternal daylight. At Baraboo, special lights go on at dawn and dusk to trick the birds into thinking they are much further north than Wisconsin. Improved egg production proves the subterfuge is working.

George Archibald acknowledges that the cranes'

Right: *Siberian cranes, bred successfully at George Archibald's American crane sanctuary, have already produced eggs which have been returned to the USSR. They were, in fact, placed in common crane nests which he hopes will confine the new race of Siberian cranes to safer territory.*

dignity may be suffering somewhat, but regards their endangered status as being more important. Fortunately the determined approach works. The bird mothers of Baraboo are thoroughly dedicated to their work. It gives them the results they believe are needed if the world is to keep its rare cranes.

Every serious newspaper in the world carried the news in 1981 when the first Siberian crane chick was hatched in captivity. The bird was called Dushenka and she hatched at the ICF, Baraboo.

Siberian cranes and politics
I.C.F.'s long-term ambition is to throw a protective net over all the world's cranes. This means acquiring or breeding 15 genetically unrelated pairs of each of the endangered species and subspecies. They have to find room and board for 350 birds!

Right at the start of this project, Archibald and Sauey realized that to rescue a bird which regards the world 'as its oyster' would involve new levels of international diplomacy to rival a United Nations agency.

'It's not just a matter of distances,' George Archibald points out. 'The cranes traverse very tense political boundaries. For example the Siberian crane is hunted extensively when it migrates into Afghanistan and India. They are large white birds with black wings, a very tempting target: they are probably edible too.'

The very existence of the ICF was enough to make a major difference. George Archibald is being perfectly accurate when he says that until he and his colleagues began intensive work on the Siberian crane, nobody knew very much about the bird, even though a number of Russian conservationists had expressed concern about its dwindling numbers. Siberian cranes are thought to be the rarest migratory birds in Asia and are extremely endangered. At best there are 150 of them left in the entire world. Two-thirds of these live in a colony in Eastern Siberia. They could easily be the very last of their kind, as the other two smaller groups which winter in Iran and India are slowly shrinking each year.

George Archibald decided to contact the Russians after he had made a study of Siberian cranes in India, which revealed just how close they were to extinction. Starting with a letter to a scientific colleague, the relationship blossomed into mutual visits and culminated in the chick called Aeroflot! 'This is a truly

GEORGE ARCHIBALD
Cranes

unique bird,' George grinned. 'The only one of its kind or any other to be hatched at 35,000 feet. I was bringing this Siberian crane egg back from Russia in a box, wrapped in a cashmere sweater with a hot-water bottle, when it started to hatch. I had excellent co-operation from the Aeroflot hostess who came to me every 15 minutes to ask if I needed the hot water filled up. But everything worked out fine. Of the seven eggs the Soviets collected for us, we managed to hatch five.'

George has even worked out what can only be described as a devious plot for overcoming the problem of where new generations of Siberian cranes will winter when they are eventually reintroduced to the wild. 'We intend that our eggs should go back to the Soviet Union for a programme that will, hopefully, produce a new generation of Siberian cranes. But we are still stuck with the problem of protecting them outside the native borders.

'We are going to try putting these eggs into the nest of another species, the common crane. The female won't know the difference between its own and the Siberian crane egg, and when the Siberian chick hatches she will lead it and her own chick to the common crane's wintering ground. That would shorten the migration route of these new generations of Siberians and confine it to safer territory.'

Author's Note: This has now been done. Four ICF eggs were transported safely to Russia; unfortunately only one proved to be fertile and, while it is believed the egg was incubated, it is not known whether it hatched: this unique crane 'cuckoo' has yet to be spotted on common crane wintering grounds.

Red-crowned crane
By concentrating on a particular family of birds, ICF ornithologists have acquired unique and invaluable knowledge of cranes, and these days George Archibald spends a lot of time on the road advising conservationists from other countries. He makes regular visits to Japan where the fate of the red-crowned crane represents a classic case of Man and animal in conflict, even in a country where cranes are especially revered. Cranes mate for life, and the Japanese Shinto religion respects them as a powerful symbol of longevity and marital fidelity.

Nonetheless, the Japanese industrial boom keeps the breeding wetlands of the magnificent red-crowned cranes under constant pressure from the adjoining paper-making complexes of Kushiro. The problem is compounded by the fact that although there are migratory populations of Japanese cranes wintering in Korea and Eastern China, the flock breeding near Kushiro is non-migratory and therefore totally dependent on that habitat.

George Archibald's most recent visits revealed that 'There are perhaps 270 birds left in Japan. They nest in southeastern Hokaido in the far north of the archipelago. It's the last area where natural wetlands are left, and each pair requires about $1\frac{1}{2}$ square miles of undisturbed shallow open wetland for successful breeding. They are highly territorial birds, and need that much space in order to rear their young. Unfortunately, many of these wetlands are now being reclaimed for agriculture and industry. There is also a tunnel under construction between Honcho and Hokaido and when that is completed, human access

GEORGE ARCHIBALD
Cranes

Below: *The Japanese shinto religion regards the country's red-crowned cranes (who mate for life) as powerful symbols of longevity and fidelity. Nonetheless Japanese cranes share the uneasy fate of cranes worldwide, as the country's industrial revolution eats up more and more wild habitat.*

GEORGE ARCHIBALD
Cranes

Above: *An apt symbol of powerful, graceful flight, Japan Air Lines chose the crane as their logo: the irony is that half the world's cranes are so close to extinction they may not be flying by the turn of the century – and that could well include Japan's red-crowned cranes.*

to Hokaido will be much easier and the pressure for development will be that much greater.'

The ICF and its Japanese colleagues have been working on wetland conservation in Japan since 1972, but in spite of the bird's very special status in Japan (it is even the symbol, denoting safety, of Japan Air Lines), George Archibald is far from sanguine about the cranes' future. So back at Baraboo they are working on the establishment of a breeding population of Japanese cranes as a matter of urgency. George Archibald now makes regular visits to China where crane consciousness is growing. He has a plan for boosting the population of red-crowned cranes native to China by 'planting' captive-bred birds in their winter roosts.

'There are about 150 of these very threatened cranes wintering in the Korean demilitiarized zone, but over the last two or three winters there have only been about two or three chicks. It's an ageing population, and we think the lack of breeding is the result of the development of the wetlands along the Sino-Soviet border. The cranes simply can't breed. Something has to be done to help this group stay alive until the Chinese can set up wetland sanctuaries. We think we should release some of our captive-bred young birds in these wintering grounds.

'Cranes actually learn their migration route from their parents by doing a round trip together in the spring. We hope our birds will pair with the wild ones, and migrate with them to learn and sustain the knowledge of where and how to migrate.'

Sandhill crane
Another case is the native crane of Wisconsin, the sandhill crane, living in the reed beds near Baraboo. This crane is particularly fascinating because there are fossil traces of it dating back nine million years. That makes it the world's oldest-surviving bird species.

But 40 years ago the sandhill was very nearly obliterated. Hunting and habitat attrition (in particular the draining of wetlands during the American depression) had reduced the cranes to about 70 birds. American conservation groups decided that here was a tragedy that could be averted, and they tackled it with real old-fashioned American 'get-up-and-go'. Archibald's ICF were enthusiastic participants in the final stages of this dramatic rescue.

It was decided that the future of the sandhills could only be assured if they had absolute rights on the wet-

lands which are vital to their lifestyle. Teams of bird-
watchers tracked down each and every stopover used
by the sandhills in Wisconsin, and with the co-
operation of the State Government, these were desig-
nated crane sanctuaries.

Now, to everyone's delight, not least of all the team
at ICF who see it as a vital precedent for rescues
elsewhere in the world, the sandhill population is
booming.

'It's a success story!' George Archibald quotes with
jubilation. 'No, it's better than that, it's a good luck
story.'

New crane countries?

If George Archibald and the ICF team offer hope for
birds that are as endangered as the red-crowned and
Siberian cranes, is there a still larger prospect? Could
those countries, that have already lost their crane
populations, consider re-establishing them with ICF
stock and techniques? That surely would be the
miracle of a century where we have come to accept
that once a country loses its wildlife, it is gone for
good.

George Archibald believes such a rebuilding programme is both possible and desirable, and he has
held talks with Sir Peter Scott of the Wildfowl Trust
about the common crane. Common by name it may be,
but as the picture on page 212 shows, it is a magnificent bird that would considerably enrich British bird-
watching. In fact, it would become our largest bird.
It is a species that was historically resident in Britain
and once bred here. It vanished from these islands
before we had knowledge of migration routes, but
modern ornithologists believe it probably wintered
in North Africa; attrition along the migration route
was the most likely cause of its demise. But is it
actually possible to cause long-vanished species to
reside again in their former homes?

'The new populations would probably be non-
migratory,' George Archibald suggests. The ICF believes in keeping its cranes where they will be safe.
'They would probably have to be fed artificially in
winter, but this particular species can tolerate any
extreme, as long as there is some kind of food source.'

This question of the degree to which we should
manipulate the course of nature is under very heavy
debate. The purists feel that a new British crane of
the type proposed by George Archibald *et al* is a
humanized copy. If it does not migrate, is it really any-

GEORGE ARCHIBALD
Cranes

thing other than a caged bird, in effect a zoo animal, admittedly in a very large zoo?

The question is also a thorny one for conservationists and their support agencies. Should they spend their limited funds on creatures with a genuine chance of survival, or on the crisis cases? This could be the most important decision nature-watchers of our day will have to make, and one that has to be made quite soon. Hundreds of species are verging on extinction, and present policies simply don't work.

Having had personal contact with a number of these cases, we favour any policy that does work. It may be that we need a global rationale for wildlife, but it does not exist yet and until it does we will back the fire-fighters like George Archibald. His attitude is unequivocal: 'I don't think some of these birds are going to last to the end of this century. Without some kind of management, active management, by man, the Siberian crane will disappear from the wild.'

Crane postscript

* In 1984, a 22,000-hectare sanctuary was created by the Chinese government in Jiangxi Province to protect the largest flock of Siberian cranes in the world, thus ensuring that two of the three wintering areas for Siberian cranes are now protected.
* In September 1982, George Archibald was awarded conservation's highest honour, the Order of the Golden Ark, by Prince Bernhard of the Netherlands – 'for his outstanding efforts in many countries on behalf of the world's endangered cranes, and in particular for his tireless work in building up the International Crane Foundation'.
* Pakistan has conducted a survey and launched a conservation initiative to stem the hunting of migrating cranes by Pathan tribesmen.
* In August 1982, India's first sarus crane count was launched with the objective of 'involving local people in the preservation of rare species, and to make them more aware of the cause of nature conservation'.
* 800 Siberian cranes have been found wintering in Jiangxi Province in late 1984.

MAN

18. THE LAST WATCH

Domination by a species has always been a precursor of extinction of other creatures; this is the problem that faces the human race. A solution may lie in our understanding that we must share our living space on this planet with other animals. But it is a lesson we, like all dominant species before us, have always had difficulty in recognizing.

In 1984, we were presented with a unique opportunity to investigate the vital, essentially political, issue of conversation. It was as if the location and the players had been pre-arranged for our purpose. Ngorongoro, a volcanic crater in the heart of Tanzania, is an animal kingdom that has been described as the eighth wonder of the world. A few miles away had lived the first man-ape, *Zinjanthropus*. Today his place has been taken by one of Africa's last remaining great tribes, the Maasai.

One of black Africa's most experienced and coherent conservationists, Solomon ole Saibull, the conservator of the Ngorongoro Crater, told us about the Ngorongoro experiment of peaceful human–animal cohabitation.

Above: *Solomon ole Saibull, Conservator of Tanzania's Ngorongoro Crater (below).*

SOLOMON OLE SAIBULL:
A sad lesson

Saibull has grappled at every political level with all the most vital conservation questions, not least the future of wildlife in the context of African nationalism. He is also a proud Maasai, respectful of his tribe's long history, culture and traditions.

He is better placed than any other man we have met to make the judgement of whether it is possible for the people, who are his family, and the wildlife, to which he has devoted his life, to share the same space.

The Ngorongoro Crater

There are very few places on Earth to rival the beauty of the Ngorongoro Crater. It is a cornucopia of wildlife, thrust up by volcanic activity out of the dry plains of Africa, and now surrounded by a ghostly patch of mountain forest draped in Spanish moss.

'Very few areas in the world are left with such a variety of animals in such large numbers,' Solomon ole

Opposite: *Maasai – one half of an impossible choice.*

SOLOMON OLE SAIBULL
The Ngorongoro Crater

Saibull admitted, aware that he was being modest. 'It is different from the rest of Africa; indeed, from the rest of the world.'

The crater is ten miles wide, with steep walls and a typical crater lake at its centre. Several similar craters exist in other parts of the world, indeed Tanzania itself has several. But the Ngorongoro Crater has 25,000 grazing wild animals and their associated predators packed into it!

'All the big cats can be found on the floor of the crater, including leopards and a good population of cheetahs,' Saibull confirmed. 'The lion population (bear in mind the very small area) is rarely less than 80, and has reached 120. There are about 100 rhino in and around the crater, although their numbers have been reduced by poaching, and it is hard to be absolutely sure how many there are.

'The populations of traditional plains game like wildebeeste and the other antelopes are substantial. There have also been some interesting additions since the place was first surveyed by European explorers 100 years ago. Then there were no reports of any buffalo, but now we have at least 2000 in the crater. It is much the same with elephants. Sometimes, when the crater trees have been bashed down, there are none, while at other times there may be as many as 40.'

Remember that we are talking of a piece of land which can be driven across in less than half an hour!

The Ngorongoro reserve is established
Saibull first saw Ngorongoro as a young student deeply committed to the African nationalist movement. 'At the time there was a controversy between the Maasai and the Colonial Administration, who wanted to protect Ngorongoro as a park. My reaction at that time was that protecting an area for animals rather than for people was wrong. That, coupled with the fact that the quarrel was between the Colonial Administration and my own people, put me definitely on the side against the park.'

The controversy smouldered on. Saibull was at university when an official committee of enquiry came out from Britain to resolve what had become known as 'The Serengeti Question'. Five years later a compromise was reached which Saibull, now a rising civil administrator, thought would be a 'lasting solution'.

'It was agreed that the animals would be protected,

but that the Maasai would continue to live in the conservation area, which encompassed Ngorongoro, at peace with the animals.'

So the stage seemed set for an ideal experiment in cohabitation, a test that conservationists badly needed if they were to resolve similar problems of confrontation in other parts of the world. Initially, all the signs pointed to success.

With their nomadic lifestyle and mobile food supply, the Maasai would not farm the bush and kill the game that invaded their crops, as other tribes were doing all over Africa. Nor would they set up permanent houses that would grow into villages, towns and cities, ousting wildlife. Finally, the experiment was to be monitored by Solomon ole Saibull, a Maasai who loved his country and its wildlife heritage but, as a nationalist, was well aware of the political realities.

This young man then found himself in an extraordinary position. Tanzania became independent

SOLOMON OLE SAIBULL
The Ngorongoro Crater

Left: *The once-prolific rhino has been decimated by human avarice. Their horns of impacted hair are avidly sought as aphrodisiacs and dagger handles by rich potentates. Solomon ole Saibull has put the protection of Ngorongoro's rhinos on a war footing, and has already had to fight off a helicopter gunship.*

SOLOMON OLE SAIBULL
The Ngorongoro Crater

from Britain, and Saibull was offered the job of conservator of the huge conservation area; he was to take over from white game administrators who were legends in their field, who had ruled the Serengeti and Ngorongoro with a feudal autocracy which the naturally well-disciplined Maasai had resented, but respected.

'I came here to take over about ten years after I had first visited the place as a boy,' Solomon recalls. 'I immediately found it very attractive and started to get interested in wildlife and conservation. At that time I had no background at all in these things. I read everything I could find on the Ngorongoro and started spending a lot of time in the crater itself. That interest has just gone on and on; even today I can stay in the crater all day and still find something new.'

Since then, still very much a Maasai and with the added experience of a spell as a government minister, he has faced up to the question of the cohabitation of man and animals. The process has been agonizing. Saibull made the protection of the parks (in which he fervently believes) his first priority, even though this involved him in war: bush fighters, the Somali 'shufta' from Kenya, swept into the Serengeti and the Ngorongoro conservation area in the late 1970s.

'It became necessary to station a special highly mobile force on the floor of the crater, to keep track of these poachers and to keep a special eye on the rhinos. It is a war which will continue,' Solomon ole Saibull accepted philosophically.

He has come to regard the Ngorongoro Crater as a fortress, one which he will make impenetrable.

This bizarre war, however, is the least of Solomon ole Saibull's problems.

'The crater really is like a fortress. It is not easy for any poacher to enter undetected. However, if the killer is inside the fort, then you are in real trouble.'

The Maasai question

As the years passed, Saibull began to suspect that the compromise calculated to solve the Serengeti question was not going to work, because of the Maasai. They are renowned for their fierce independence.

It used to be said that they were the only major tribe in Africa who were never colonialized. Both the German and British rulers of Tanganyika (Tanzania) more or less left the tribe to go its own nomadic way. Even the missionaries made little impact, and the fabric of Maasai culture has remained largely intact.

Both their independent nature and their culture are the product of the Maasai's singular lifestyle. They are nomadic cattle herders living on a highly nutritious diet composed largely of blood and milk, both taken from live cattle, giving the Maasai an assured supply of food, and great mobility. Their traditions tell them that only they are good at looking after cattle; in fact they were put on Earth for that very purpose. Everyone else who has cattle is therefore operating in defiance of divine purpose and should be relieved of their cattle as soon as possible.

There were also some benefits for the rest of society in the Maasai way of life. Thanks to their cattle the Maasai did not need to kill game for food (other than when an animal attacked their cattle). Furthermore, the Maasai homeland is located to the north of the Serengeti National Park, and has always been something of a no-go area for intruding poachers from Kenya.

SOLOMON OLE SAIBULL
The Ngorongoro Crater

Below: *Once as nomadic as the animals around them, the Maasai life style is changing. Manyattas grow more permanent and the farming of crops is becoming a new occupation. This all adds up to less land that can be shared with the wildlife.*

SOLOMON OLE SAIBULL
The Ngorongoro Crater

The only wild creatures that had anything to fear from the tall, athletic, spear-carrying Maasai were lions. Killing a lion with a spear was part of the initiation to manhood. Conservationists believe that the Maasai eliminated the practice years ago, but we were given to understand that it might still exist in the remote hinterland. Maasai are fiercely protective of their traditions and still practise the ritual of full circumcision of both male and female teenagers, so the report could well be true.

There was a brief period around the time of independence when the Maasai started to kill game, but Solomon ole Saibull, who had to deal with it, believes this was essentially political. 'It happened when the decision was taken to set up the conservation area to protect the wildlife in areas where the Maasai lived. Their reaction was to turn on the animals that were now limiting their rights within the area. It didn't last long.'

Why the experiment failed
'We, like everyone else, have failed to find a solution to the human need for development,' Saibull told us. 'We were all being romantic. The Maasai were bound to develop; their needs would change, their relationships with the land, their cattle, and the animals, would alter.'

And they did change, until eventually Saibull had to face the harrowing (and dangerous) task of evicting his own people from Ngorongoro and resettling them in areas where, in effect, the animals would take second place. 'The major change was in their food habits. The Maasai used to rely entirely on their herds for food. Now they are depending more and more on grain, which requires the cultivation of land. The way of life is also changing. More and more Maasai are being educated, getting used to living in larger and more modern houses, neither of which are compatible with a conservation area.'

Saibull believes that very soon the nomadic life will also disappear; in fact, in the light of the other changes, 'responsible' government policy is to encourage the process. 'Otherwise the Maasai will be left behind in the stream of social development, and end up like the reservation Indians of North America – a pathetic state of affairs that nobody would wish on them.'

The end of the experiment

Having reluctantly accepted that the experiment had been a dream – 'that the Maasai would have to be frozen in time in order to remain part of the landscape' – Saibull embarked on his second local war, to evict the Maasai from the Ngorongoro and Empakai Craters.

This was the most difficult battle. Saibull saw a state of total decay looming before the Maasai, while they saw him as betraying his tribal loyalties, the ultimate treachery.

The Ngorongoro move was conducted without real resistance. Saibull believes the Ngorongoro Maasai had grown tired of the daily tourists 'relegating them to the role of the animals they lived among!' When he approached the seven families living in greater isolation in the second crater it was a different story.

'My father had died the year before and one of the elders came up to me and asked: "Where is your father?"

'I replied: "He is dead".

' "We know," the elder told me; "And this should be the last time you come to this crater, otherwise we will make sure you follow your father."

'At the same time this man had started to shake like a warrior, and the ranger who was with me cocked his rifle! I asked the elder: "If you, an elder, are going to shake with rage, what should the warrior (my ranger) do now?" The elder said "He may do as he pleases."

'I realized then that the situation was not good and I told him "Warriors obey their elders", which seemed to calm him down. He repeated that I should not come back to the crater and I told him he had until Sunday to leave. We made a kind of gentlemen's agreement and they let me go.'

It did not work! Saibull himself led a small army detachment the following Sunday and they camped overnight on the crater rim, allowing the elder to keep his promise.

'The following morning the Maasai were still there and I decided to scare them out using flares and thunder-flashes. Some of them started to move out but the die-hards just sat tight and armed themselves.

'We went down onto the crater floor and one of them came at me with a panga (a very heavy knife) but one of the sergeants ordered him to "stop his arm", and thankfully he did! Finally we managed to disarm them without any casualties. We took the more serious weapons such as spears and swords but

SOLOMON OLE SAIBULL
The Ngorongoro Crater

Below: *Maasai warriors. Can man and animals share the same wild habitat? The answer would appear to be no; and in the case of these unique Tanzanian crater sanctuaries, priority was given to the irreplacable wildlife resource.*

we gave them back their sticks. It's very bad to disarm a Maasai and take away everything.'

Worlds apart

Our experiences with the Maasai and Solomon ole Saibull have brought us to the reluctant conclusion that, for the time being at least, and perhaps for the foreseeable future, man must go into voluntary exile from certain wild places, setting them aside for the sole use of animals. The only human access should be that which helps finance these sanctuaries, or visits by nature-watchers whose studies improve the lot of the animals.

Solomon ole Saibull agrees with this more practical ideal and is about to embark on his third major campaign – the reversion of the entire Ngorongoro conservation area to animal use only, which will involve resettling even more Maasai families.

He is well aware how difficult this could be.

'It is very hard for any human being to maintain

the idea that people rather than animals should disappear. But this is not really the point. Animals are not preserved for their own sake but for man. Even when left to themselves, their fate is still in the hands of the people.'

Nor does he believe that the affluent West, whose tourists get the most use from the game parks, should foot the bill for wildlife conservation, as we have heard so frequently suggested elsewhere in Africa. This still-proud Maasai presented us with an enlightened overview – a new philosophy for world conservation, perhaps?

'It is something that poor countries, more than any others, must be able to afford. Rich countries like the United States have enormous economic power. If it should make mistakes about its nature areas it can correct them. For a poor country it is essential to make sure that we don't make mistakes. Our wildlife is a valuable resource, one that can never be replaced.'

SOLOMON OLE SAIBULL
The Ngorongoro Crater

RECOMMENDED READING

NEVILE COLEMAN

What Shell is That?
by N. Coleman (Lansdowne, Australia)
Australian Beachcomber
by N. Coleman (Collins)
A Field Guide to Australian Marine Life
by N. Coleman (Rigby, Australia)
Australian Sea Fishes 30° South
by N. Coleman (Doubleday)
Australian Sea Fishes 30° North
by N. Coleman (Doubleday)

EUGENIE CLARK

Red Sea Reef Fishes
by John E. Randall (Immel Publishing Co., London, 1983)
Mysteries of the Red Sea
by Lev Fiskelson (Massada Ltd., Givatayim, Israel, 1984)
Marine Life
by David & Jennifer George (Harrap, London, 1979)
Saudi Arabian Seashells
by Doreen Sharabati (Royal Smeets Offset, B.V. Weert, 1981)
Biotopes of the Western Arabian Gulf
by P. W. Basson, J. E. Burchard, J. T. Hardy & A. Price (Aramco, New York, 1977)
The Book of Sharks
Richard Ellis (Harcourt Brace Jovanovich Publishers, 1983)
Red Sea Diver's Guide
by Shlomo Cohen (1975)

STRUAN SUTHERLAND

Venomous Creatures of Australia
by Struan K. Sutherland (Oxford University Press, Melbourne, 1982)
Australian Animal Toxins
by Struan K. Sutherland (Oxford University Press, Melbourne, 1983)
Take Care! Poisonous Australian Animals
by Struan K. Sutherland (Hyland House, Melbourne, 1983)

JOHN E. RANDALL

Studies in Tropical Oceanography
by J. E. Randall (University of Miami, Institute of Marine Science, 1967)
Sea Frontiers
by J. E. Randall (1971)
Caribbean Reef Fishes
by J. E. Randall (T.F.H. Publications, Redhill, Surrey, 1983)

Red Sea Reef Fishes
by J. E. Randall (Immel Publishing, London, 1983)
Hawaiian Reef Fishes
by J. E. Randall (Harrowood Books, Newton Square, Penn)

ALAN MITCHELL

The Trees of Britain and Northern Europe
by Alan Mitchell & John Wilkinson (Collins, 1982)
Identification of Trees & Shrubs
by E. Makins (1940 approx)
Trees & Shrubs Hardy in the British Isles
bu W. J. Bean (Revised by Desmond Clarke)
Hillier's Manual
Trees of Great Britain & Ireland
by Elwes & Henry (1904–1912, 7 volumes)
Manual of Cultivated Trees
(1940)

TONY HALL

Conservation of Threatened Natural Habitats
by A. V. Hall (South African National Scientific Programmes Report, CSIR, Box 395, Pretoria, 1984)
Threatened Plants of Southern Africa
by A. V. Hall, B. de Winter & S. A. M. van Oosterhout (South African National Programmes Report No. 45, CSIR as above, 1980)
Threatened Plants of the Cape Peninsula
by A. V. Hall & E. R. Ashton (Threatened-Plants Research Group on behalf of the Cape Peninsula Conservation Trust, Box 4637, Cape Town, 1983)
Endangered Species in a Rising Tide of Human Population Growth
by A. V. Hall (Transactions of the Royal Society of South Africa, Volume 43 Part 1, Royal Society of South Africa, University of Cape Town, 1978)
Wild Orchids of Southern Africa
by J. Stewart, H. P. Linder, E. A. Schelpe and A. V. Hall (Macmillan South Africa, Johannesburg, 1982)

JOHN SIMMONS

Life of Plants
by John Simmons (Macdonald Educational, 1974)
Royal Botanic Gardens, Kew (Gardens for Science & Pleasure)
edited by F. N. Hepper (Her Majesty's Stationery Office, 1982)
Conservation of Threatened Plants
edited by J. B. Simmons *et al* (Plenum Press, New York, 1976)
Survival or Extinction
edited by Synge and Townsend (Bentham-Moxon Trust, R.B.G., Kew, 1979)

Creating Habitats for Living Collections
by J. B. E. Simmons (R.B.G., Kew, 1983)

TOM EISNER
Sociobiology
by E. O. Wilson (The Belknap Press, Harvard University, Cambridge, 1975)
The Insect Societies
by E. O. Wilson (The Belknap Press, 1971)
The Dance Language & Orientation of Bees
by K. von Frisch (The Belknap Press)

MIKE AUGEE
The Biology of Monotremes
by M. E. Griffiths (Academic Press, New York, 1978)
The Platypus
by T. Grant (New South Wales University Press, Sydney, 1984)
Monotremes and Marsupials: the Other Animals
by T. J. Dawson (Edward Arnold, London, 1983)
Monotreme Biology
edited by M. L. Augee (Royal Zoological Society of N.S.W., Sydney)
Echidnas
by M. Griffiths (Pergamon Press, London, 1968)

JONATHAN SCOTT
Pyramids of Life
by John Reader and Harvey Croze (Collins)
Portraits in the Wild
by Cynthia Moss (Hamish Hamilton, 1976)
The Serengeti Lion
by George Schaller (University of Chicago Press, Chicago, 1972)
The Long African Day
by N. Myers (Macmillan Publishing Co. Inc., New York, 1972)
The Marsh Lions
by Jonathan Scott and Brian Jackman (Elm Tree Books, 1982)

DAPHNE SHELDRICK
The Maneaters of Tsavo
by Colonel Patterson
Claws
by Colonel C. A. Brown
Jock of the Bushveldt
by Sir Percy Fitzpatrick
A Game Rangers' Notebook and
A Game Ranger on Safari
by Blaney Percival
A Kenya Diary
by Meinetzhagen

ANNE RASA
Mongoose Watch: A Family Observed
by Anne Rasa (John Murray, 1985)
Introduction to Animal Behaviour

by Aubrey Manning (Addison-Wesley Publ. Co., Reading, Mass., 1967)
King Solomon's Ring
by Konrad Lorenz (Methuen, London, 1952)
Sociobiology
by E. O. Wilson (Belknap Press, London, 1975)
Animal Behaviour
by R. Hinde (McGraw-Hill, New York, 1970)
The Study of Instinct
by N. Tinbergen (Clarenden Press, Oxford University, 1951)

ALISON JOLLY
Social Behaviour in Primates
by Neil Chalmers (Edward Arnold, London, 1979, Open University text book)
The Woman that Never Evolved
(Harvard University Press, London, 1981)
A World Like Our Own
by Alison Jolly (Yale University Press, London, 1980)
The Evolution of Primate Behaviour (2nd edition)
by Alison Jolly (Macmillan, New York, 1984)
Madagascar (Key Environments Series)
edited by Alison Jolly, Philippe Oberlè & Roland Albignac (Pergamon Press, Oxford, 1984)

ROGER WILSON
Gorillas in the Mist
by Dian Fossey
The Mountain Gorilla
by George Schaller (Chicago University Press)
The Natural History of the Gorilla
by Dixon
Gorilla Behaviour
by Terry L. Maple and Michael P. Hoff (Van Nostrand Reinhold)
The Wandering Gorillas
by A. G. Goodall (Collins, London)

BILL ODDIE
Discovering Birds
by D. I. M. Wallace
British Birds
(Monthly magazine)
The Shell Guide to Birds of Britain and Ireland
by James Ferguson-Lees, Ian Willis and J. T. R. Sharrock (Michael Joseph)
Field Guide to the Birds of North America
(National Geographic)
The Birds of the Western Palearctic (Vols 1 to 3)
by Cramp & Simmons (Oxford University Press/RSPB)

ROBIN BAKER
The Mystery of Migration
by Robin Baker (Macdonald, 1980)
Human Navigation and the Sixth Sense
by Robin Baker (Hodder & Stoughton, 1981)

Migration: Paths through Time and Space
by Robin Baker (Hodder & Stoughton, 1982)
Bird Navigation: the Solution of a Mystery?
by Robin Baker (Hodder & Stoughton, 1984)

GEORGE ARCHIBALD
Cranes of the World
by P. Johnsgard (Indiana University Press, Blooming-
ton, Indiana, 1983)
Crane Research Around the World
by J. Lewis (International Crane Foundation, Route 1,
Box 230C, Baraboo, Wisconsin 53913, USA, 1981)
The Whooping Crane
by R. P. Allen (Monograph Series, National Audubon
Society, 1000 Fifth Avenue, New York, 1952)
Cranes of the World
by L. H. Walkinshaw (Winchester Press, 460 Park
Avenue, NY 10022, 1973)
The Japanese Crane – Bird of Happiness
by D. Britton (Kodansha International, 12–21 Otowa
2-chome, Bunkyo-ku, Tokyo 112, Japan, 1981)

SOLOMON OLE SAIBULL
Herd and Spear: The Maasai of East Africa
by Solomon ole Saibull and Rachel Carr (Collins, 1981)

Ngorongoro: The Eighth Wonder
by Fossbrooke (Andre Deutsch, 1972)
First Visitor, Geological History, Trees and Shrubs
(Ngorongoro booklets)

CARL JONES
Year of the Greylag Goose
by Konrad Lorenz
The Falcons of the World
Tom Cade and David Digby (Collins)
The Goshawk
T. H. White (Penguin)

ROGER PAYNE
The Ecology of Whales and Dolphins
D. E. Gaskin (Heinemann 1982)
The Sierra Club Handbook of Whales and Dolphins
Leatherwood, Reeves and Foster (Sierra Club Books,
1983)
The Book of Whales
R. Ellis (Alfred A. Knopf, 1980)
*Marine Mammals of Eastern North Pacific and Arctic
Waters*
Edited by D. Haley (Pacific Search Press, Washington,
1978)

ACKNOWLEDGEMENTS

Picture research: Bryony Kinnear

The publishers would like to thank Jillian Luff for the maps on pages 15, 18, 23, 29, 37, 47, 66, 76, 79, 85, 93, 99, 104, 111, 121, 133, 143, 155, 161, 168, 182, 185, 190, 203, 213, 225: and the following for permission to use their photographs: Page 8 Robin Brown; 9 Central Television; 12 Lawson Wood; 15 Neville Coleman; 16 Neville Coleman; 18 Bryony Kinnear; 19 Dr Eugenie Clark © 1981 National Geographic Society; 21 David Doubilet; 23 Dr Struan K. Sutherland; 24 Neville Coleman; 25 Neville Coleman; 26 P. J. Probert; 28 Neville Coleman; 29 Ben Patnoi; 31 Neville Coleman; 32 Neville Coleman; 34 Neville Coleman; 37 Bryony Kinnear; 39 Doninick Macan; 40 David Doubilet; 42 Dr Horace E. Dobbs; 43 Lawson Wood; 44 Lawson Wood; 46 Flip Nicklin; 47 Bryony Kinnear; 50 Ken Balcombe, Bruce Coleman Ltd; 55 Lysa Leland; 57 Dr Pieter Lagendijk; 58 Dr Roger Payne; 59 WWF/Hal Whitehead, Bruce Coleman Ltd; 62 Dr Ronald Stecker; 65 Dr Ronald Stecker; 66 Dr H. Thomas Harvey; 68 Dr Ronald Stecker; 69 Dr Ronald Stecker; 71 Dr Ronald Stecker; 72 Dr Ronald Stecker; 74 Bryony Kinnear; 75 Alan Mitchell; 76 Alan Mitchell; 78 Prof A. V. Hall; 79 Robin Brown; 80 Prof A. V. Hall; 81 Robin Brown; 82 Prof A. V. Hall; 85 John Simmons; 86 John Simmons; 88 John Simmons; 90 Royal Botanic Gardens, Kew; 91 Royal Botanic Gardens, Kew; 92 Dr Tom Eisner; 93 Julian Pettifer; 95 F. Schauff; 98 Julian Pettifer; 99 Robin Brown; 100 Dr Tom Eisner; 102 Dr Tom Eisner and D. Aneshansley; 104 Bryony Kinnear; 105 Dr Struan K. Sutherland; 107 Dr Struan K. Sutherland; 110 Bryony Kinnear; 112 Elma Garrick; 115 Bryony Kinnear; 117 Tom Grant; 118 Tom Grant; 120 Jonathan P. Scott; 121 Robin Brown; 123 Jonathan P. Scott; 124 Jonathan P. Scott; 126 Jonathan P. Scott; 127 Jonathan P. Scott; 130 Jonathan P. Scott; 132 Daphne Sheldrick; 133 Robin Brown; 136 Daphne Sheldrick; 138 Robin Brown; 139 Daphne Sheldrick; 142 Dr Anne Rasa; 143 Bryony Kinnear; 146 Dr Anne Rasa; 149 Dr Anne Rasa; 150 Dr Anne Rasa; 152 Bryony Kinnear; 154 Bryony Kinnear; 155 Bryony Kinnear; 156 Bryony Kinnear; 157 Mark Pidgeon; 160 Bryony Kinnear; 161 Karen Stanley-Price; 163 Robin Brown; 164 Robin Brown; 166 Roger Wilson; 168 Robin Brown; 170 Roger Wilson; 173 Bryony Kinnear; 175 Bryony Kinnear; 176 Bryony Kinnear; 178 Roger Wilson; 180 M. W. Richards, RSPB; 182 Leitz Instruments; 184 Oddsocks Press/Eyre Methuen (cartoons); G. G. Bates, Aquila Photographics; 185 Bobby Tulloch; 187 Bobby Tulloch; 188 Jim Clare; 190 Central Television; 192 W. S. Paton, Aquila Phorographics; 193 Edgar T. Jones, Aquila Photo-graphics; 195 G. W. Ward, Aquila Photographics; 196 Philip Shaw, Aquila Photographics; 198 Dr Robin Baker; 202 Carl Jones; 203 Bryony Kinnear; 204 Bruce Coleman Ltd; 205 Carl Jones; 206 Carl Jones; 210 Carl Jones; 212 Dr George Archibald; 213 Dr George Archibald; 216 Dr George Archibald; 219 Dr George Archibald; 220 Japan Air Lines; 224 Bryony Kinnear; 225 Bryony Kinnear; 227 Bryony Kinnear; 229 Bryony Kinnear; 232 Bryony Kinnear.

INDEX